Praise for *Transformer*

'Nick Lane's exploration of the building blocks that underlie life's big fundamental questions – the origin of life itself, ageing and disease – have shaped my thinking since I first came across his work. He is one of my favourite science writers.'

Bill Gates

'In this compulsively readable book, Lane takes us on a riveting journey, ranging from the flow of energy to new ways of understanding cancer. Lane provides a luminous understanding of how scientists, including Lane himself, are rethinking energy and living organisms.'

Siddhartha Mukherjee, author of
The Gene: An Intimate History

'Thrilling and highly persuasive ... This hugely important book is set to become a landmark, transforming our understanding of how life works.'

Gaia Vince, author of *Nomad Century*

'I loved every page of Nick Lane's new book. It's one of the very best books on the origin of life I've read.'

Lee Smolin, author of *Einstein's Unfinished Revolution*

'Hugely important ... a powerfully persuasive case for life being about energy flow, flux and change. In *Transformer*, chemistry is quite literally brought to life.'

Jim Al-Khalili, author of *The World According to Physics*

'Amazing! Takes science writing to a new level ... with soaring prose but uncompromising on scientific detail, *Transformer* made me think about life on earth in a completely different way.'

Daniel M. Davis, author of *The Secret Body*

'Hugely ambitious and tremendously exciting ...
Transformer shows how a molecular dance from the dawn of time still sculpts our lives today. I read with rapt attention.'

Olivia Judson, author of
Dr Tatiana's Sex Advice to All Creation

'Nobody explains the inner secrets of the living cell better than Nick Lane ... a series of riveting detective stories.'

Richard Fortey, author of *Trilobite!*

'An exhilarating account of the biophysics of life, stretching from the first stirrings of living matter to the psychology of consciousness. I felt as if I was there, every step of the way.'

Mark Solms, author of *The Hidden Spring*

'Nick Lane's marvellously engaging *Transformer* refocused my astronomer's gaze on the vital chemistry of life on our own planet. Both a scientific adventure story and an original quest to understand life on Earth, *Transformer* also guides us on how to find life beyond.'

John Grunsfeld, former NASA chief scientist and astronaut

'Fascinating ... Nick Lane brings together biology, chemistry and physics to illuminate the role of energy in bringing matter alive.'

Sean Carroll, author of *Something Deeply Hidden: Quantum Worlds and the Emergence of Spacetime*

'Nick Lane never writes about the living world without offering entirely new perspectives on how life itself works ... Biochemistry has never looked more exciting.'

Philip Ball, author of *Critical Mass*

TRANSFORMER

The deep chemistry of life and death

NICK LANE

P

PROFILE BOOKS

First published in Great Britain in 2022 by
Profile Books Ltd
29 Cloth Fair
London
EC1A 7JQ

www.profilebooks.com

A CIP catalogue record for this book is available from the British Library.

ISBN 978 1 78816 054 4
eISBN 978 1 78283 450 2

Typeset in Sabon by MacGuru Ltd

Printed and bound in Great Britain by Clays Ltd, Elcograf S.p.A.

In memory of
Ian Ackland-Snow

Contents

to this mortal process of continuing,
it is the movement that creates the form.

Richard Howard

LIFE ITSELF

From space it looks grey and crystalline, obliterating the blue-green colours of the living Earth. It is criss-crossed by irregular patterns and convergent striations. There's a central amorphous density, where these scratches seem lighter. This 'growth' does not look alive, although it has extended out along some lines, and there is something grasping and parasitic about it. Across the globe there are thousands of them, varying in shape and detail, but all of them grey, angular, inorganic ... spreading. Yet at night they light up, glowing in the dark sky, suddenly beautiful. Perhaps these cankers on the landscape are in some sense living – there is a controlled flow of energy, there must be information and some form of metabolism; some turnover of materials. Are they alive?

No, of course not; they are cities. We know them intimately from the inside, even if most of us know little about the flow of energy and materials through our own cities. We know them mainly by their visible structures, buildings on a map. But an empty city with no power, no energy flow, no traffic, no jostling crowds, is an eerie place, chilling and post-apocalyptic. Dead. What brings a city to life is the people, their movement from place to place, along with the flow of materials that sustains our daily existence – electricity, heat, water, gas, sewage.

It would not be misleading to say that a city is brought to life by the controlled flow of energy and material in this way. Set up a time-lapse camera on a busy street downtown and we get a sense of this flow, obeying laws of flux that we can barely guess at. In my mind's eye, we could rise above the conurbation and picture this combined flux, map out the jumbled flow of people, lights and power, pulsating down some streets, just a residual overflow in others, some districts a hive of bustling activity, others nearly dormant until the evening when the commuters return home and the lights flick on. We could map out the flux that animates a city. Certainly, we can imagine a city this way, but mostly all we notice are the buildings. The structure.

A cell is a city of a sort. It too has buildings, or at least physical structures. Unlike our own constructions a cell is not dominated by gravity, and is truly built in three dimensions. If you shrink yourself down to the size of a molecule, the 'cityscape' is dizzying. Membranes sweep past your view: curving, fluid walls, swooping overhead or plunging down below. Traffic streams past on colossal cables, extending out in all directions. Traffic as you've never seen before: great mechanical contraptions, machines the size of buildings, pistons whirring faster than the eye can see. The great citadel of the nucleus, heart of the metropolis, looms in the distance, miles away from you, but dominating your field of view. All is hustle and bustle. Unlike any human city, the vast sweeping walls themselves move and conjoin and dissociate again. Zoom out, and the whole city of the cell can change shape and move around, reassembling its internal structures as it does so. From out here, through the lens of a microscope, we can watch the trafficking of goods through the cell, lit up by fluorescent dyes like a town at night, all electric reds and blues and greens. Yet everything I have described is buildings, the structures of the cell. We can picture this marvellous city on a scale less than thousandths of a millimetre; we can visualise the movement of its parts like never before. But the cell is animated on a smaller scale still. Even the most powerful microscopes can't discriminate the moment-by-moment flow of energy and materials that animates all life, the unceasing changes that transform small molecules over millionths of seconds, and distances of less than a millionth of a millimetre. Deep within these marvellous moving structures, the flow of energy and matter is still invisible, as hard to imagine as the restless electrons that power our conurbations and the people within. Perhaps for that reason, we have a tendency to discount its importance to life.

Few things are as inscrutable as a cell. In the seventeenth century, when the Dutch microscopist Antonie van Leeuwenhoek unveiled the cosmos hidden in a drop of water, he marvelled at the little 'animalcules' that lived out their lives there, all whirring parts and purpose. For all our deep explorations of the cityscape of the cell, the behaviour of these protozoa is as beguiling and nearly as mysterious to us today. Do these microscopic blobs of animated protoplasm know what they are doing as they chase and consume each other? Surely not! But to our naive eye, it almost looks as if they do – as if these tiny beings have hopes and fears and pains of their own. As if they feel some joy or relief when they tear themselves free from the mechanical rotating jaws of some minuscule gyrating predator. Some 350 years after van Leeuwenhoek, we now know what most of these whirring parts do, what they're made of, how they function. We have taken them apart, in centrifuges or with optical tweezers, read out the code that specifies their structures, deciphered the regulatory loops that lend an illusion of purpose, listed all their parts. And yet underneath it all, we are barely any closer to understanding what breathes life into these flicks of matter. How did they first emerge from the sterile inorganic Earth? What forces coordinate their exquisite behaviour? Do they experience any sort of feelings?

For decades, biology has been dominated by information – the power of genes. The importance of genes is unquestionable, yet there is no difference in the information content of a living protozoon and one that died a moment ago. The difference between being alive or dead lies in energy flow, in the ability of cells to continually regenerate themselves from simpler building blocks.[1]

If there is a view from modern biology, it is that genetic information structures the flow of energy and materials. To a first approximation, biology is understood in terms of information networks and control systems. Even the laws of thermodynamics, which govern the behaviour of molecules and their interactions and reactions, can be recast in terms of information – Shannon entropy, the laws of bits of information. But this view generates its own paradox at the origin of life – where does all this information come from? Within the realm of biology, we already have a simple explanation: natural selection sifts through random differences, favouring what works, eliminating what doesn't, generation after generation. Information accumulates with function over time. We can quibble over details, but there is no conceptual difficulty here. At the origin of life, though, this view will not do. Place information at the heart of life, and there is a problem with the emergence

---

1 When I talk about energy flow, I'm really referring to what physicists call 'free energy', which is to say the energy available to power work (rather than being dissipated as heat). That goes for the whole book. And when I talk about building blocks, I'm referring to small molecules such as amino acids or nucleotides, which can be joined together to form giant macromolecules such as proteins or DNA, respectively. Again, that goes for the whole book.

of function, which is to say, the origin of biological information. Not only that, but there's a problem in understanding the troubling trajectory of evolution, not least the long delays between abrupt changes, such as the emergence of animals in the Cambrian explosion, despite the continuous exploration of genetic sequence space – information – across life. There are problems, too, in understanding why we age and die, why we are still suffering from diseases such as cancer, despite decades of research, and most fundamentally, how subjective experiences can give rise to the conscious mind.

Thinking about life only in terms of information is distorting. Seeking new laws of physics to explain the origin of information is to ask the wrong question, which can't be answered precisely because it is not meaningful. A far better question goes back to the formative years of biology: what processes animate cells and set them apart from inanimate matter? The idea that there is a vital force, that life is fundamentally different from inanimate matter, was disproved long ago and is now only wheeled out as a straw man to burn – even though it's an understandable illusion for anyone who has shared van Leeuwenhoek's captivation with busy animalcules. Yet biochemistry – my own discipline, which deals with the flow of energy and materials through cells – has, with a few notable exceptions, been blithely indifferent to how this unceasing flux might have arisen, or how its elemental imprint could still dictate the lives and deaths of cells today, along with the organisms they compose. You and me.

This book will explore how the flow of energy and matter structures the evolution of life and even genetic information, leaving an indelible stamp on our own lives. I want to turn the standard view upside down. Genes and information do not determine the innermost details of our lives. Rather, the unceasing flow of energy and matter through a world in perpetual disequilibrium conjures the genes themselves into existence and still determines their activity, even in our information-soaked lives. It is the movement that creates the form. I want to capture an extraordinary renaissance that is currently hiding in plain sight: how textbook biochemistry is simultaneously galvanising new paradigms on the origin of life and cancer, to name but two fields. How could such disparate questions, separated by billions of years and gulfs in planetary environment be linked? At the core of this emerging view is an amazing, conflicted cycle of reactions that uses energy to transform inorganic molecules – gases – into the building blocks of life, and the reverse. To understand this cycle of energy and matter is to resolve the deep chemical coherence of the living world, connecting the origin of life with the devastation of cancer, the first photosynthetic bacteria with our own mitochondria, the abrupt evolutionary leap to animals with sulfurous sludges, the big history of our planet with the trivial differences between ourselves, perhaps even the stream of consciousness. In this book, we will see that understanding the deep chemistry that animates life, and fades as we die, illuminates some of the enduring mysteries of biology and our own existence.

The dynamic side

To understand this flow of energy and matter and all it portends, we must first look to where biology has turned a blind eye since the ascent of information. The golden age of biochemistry began with a realisation that cells are not made of an amorphous protoplasm composed of inscrutably complex 'living' molecules. One of the founding fathers of biochemistry, Sir Frederick Gowland Hopkins, dedicated much of his long career through the first four decades of the twentieth century to promoting what he called the 'dynamic side' of biochemistry: the idea that the basic molecules of life are quite simple and can be analysed by conventional chemical methods – but that they are funnelled down specific pathways, in which one molecule is converted through some small chemical change into another form, again and again, each time fashioned by a catalyst with specific properties. Life, for Hopkins, was the combination of information, which specifies the protein catalysts (enzymes) that channel these pathways, and flux – the flow of molecules down particular pathways to form new materials for building or rebuilding the city of the cell.

I've used the word flux a few times already, and I'll use it again throughout the book. Before getting any further, let's pause for a moment to clarify exactly what I mean by it. Flux is a form of flow, but with one crucial difference. Water can flow in a river, or traffic down a street. What goes in at one end and what comes out at the other is the same thing – water, or cars. In biochemistry, flux is the flow of things that are transformed along the way. Imagine a car entering a street; let's say it's a VW Beetle. No sooner has it gone ten yards down than there's a blinding flash and it abruptly turns into a Porsche. Then another flash and it's become a Volvo. Bang! It's a white van. Zap! Now it's a minibus. Flash! It's a tractor, which leaves the street. But the strangest thing about this street is that the same thing keeps on happening: only VW Beetles ever enter the street; only tractors ever leave. The same succession of transformations takes place each time. Let's imagine that sixty VW Beetles enter the street every minute, one per second. Each of them is transformed in a series of blinding flashes into sixty tractors. That's flux: the total number of vehicles that passes down the street, each one transformed into the same type of tractor. Of course, that's just this street. Take a look at the street around the corner. There you'll see only Vespa scooters entering, transforming into Harley motorbikes. And just across town there's a canal where canoes change into speed boats.

This is the strange world of metabolic flux. Even a simple bacterial cell can undergo as many as a billion transformations per second, an incomprehensible number. You might say that there's a lot of repeat streets in a cell (with the same thing happening in each of them) but there may be several hundred vehicles entering a street *every second*, each one reliably going through exactly the same succession. This is the flux that makes up the metabolism of the cell, which we will grapple with in this book. Metabolism is what keeps us alive – it is what being alive *is* – the sum of the continuous transformations of small molecules on a timescale of nanoseconds, nanosecond after nanosecond. If we live to the age of eighty, we will have lived through nearly three billion billion (3×10^{18}) nanoseconds-worth of metabolism. No wonder we run down. We can't actually see any of this happening before our eyes, even now, but we can infer what is going on from some of the ingenious methods I'll tell you about, methods that go back to those intrepid explorers of the nanocosm, like Hopkins a century ago, who first understood that the secret of life lies in the entrained flow and rapid transformations of many plain, simple molecules.

That was before the double helix of DNA, before the information revolution in biology, before we knew much at all about how cells work. It was in fact a glorious hypothesis, based on a handful of findings in the nineteenth century – notably that some molecules of life, such as urea, could be synthesised from scratch by chemists and were in no way magic, rebutting vitalism; this was 'merely' biological chemistry, with normal chemicals that behaved according to normal rules of chemistry.

Biochemistry became the study of how these simple molecules were interconverted one into another. Did you just trip over the word 'simple'? None of this is simple, but there are levels of complexity. The molecules that I'm talking about are small, containing between one or two and up to about twenty carbon atoms, but most of them have fewer than ten carbons. Think of these as carbon 'skeletons', in which the carbon is bound to itself plus hydrogen and oxygen atoms – with, less commonly, nitrogen, sulfur or phosphorus – giving each of them their distinct properties and tendencies to react. These are the building blocks that make up cells, little more than a few hundred types of molecule in total. The giant 'macromolecules' that form the fabric of the cell, notably DNA and proteins, are actually long chains, strings of these building blocks linked together following genetic instructions that (then) remained utterly mysterious.

The glorious hypothesis that cells are animated by the continuous directed flow of simple materials and energy turned out to be true for all life. Painstaking experiments showed that the way bacteria respire with oxygen or grow in its absence is remarkably similar to how our own heart cells behave in similar circumstances. In the 1920s, another great Dutch pioneer of microbiology, Albert Kluyver, marshalled new evidence for the unity of biochemistry. His slightly maniacal phrase 'From the elephant to butyric acid bacterium – it is all the same!' (I hear it accompanied by a burst of hysterical laughter) was later famously paraphrased as 'Anything found to be true of *E. coli* must also be true of elephants.' While deliberately mischievous, there is more than a little truth in this outrageous assertion – the biochemical pathways that produce the basic building blocks of life are indeed conserved across practically all cells. Several decades later, at the dawn of molecular biology, the idea that the genetic code is universal, encoding the same twenty amino acids (the building blocks of proteins) across all life, owed much to this beguiling conception of the unity of biochemistry. The 'universal genetic code' is now a well-worn phrase, but the idea was not established rigorously at the time. It just felt right, precisely because it resonated with the unity of biochemistry.

The digital jungle

The dawn of molecular biology! It must have been intoxicating. Darwin had given order to biology a century earlier. The discovery of genes and the laws of heredity made intellectual sense of evolution. But the molecular mechanisms of inheritance remained a black box until Crick and Watson made an inspired leap to grasp the full meaning of Rosalind Franklin's beautiful, cryptic X-ray photographs of DNA (which in turn rested on Maurice Wilkins's revealing early work; it's giant's shoulders all the way down). The story that Crick burst into the Eagle pub in Cambridge and announced that he knew the secret of life is sadly apocryphal, but it has the ring of deeper truth.

The double helix of DNA is perhaps the most meaningful icon in all science. DNA is composed of two long chains of 'letters' that snake around each other into the farthest distance, each strand providing an exact template for the other. As soon as you see it, you can grasp in principle how heredity works: when the two strands are prised apart, each serves as the template to build a fresh complementary strand, giving two copies, a new double helix for each daughter. And you can grasp in principle how the genetic code works: each strand of DNA contains just four types of letter, with millions, nay billions, of these arranged in sequence down the length of the chain. Four letters might seem limited compared with, say, the twenty-six letters of the English alphabet, but the Morse code only has two letters and can convey the same meaning. We might not enjoy listening to the works of Shakespeare in blips and bleeps but technically there is no loss of meaning, and the canon can be reconstituted in full. The same goes for DNA; indeed, Shakespeare's sonnets have been encoded in synthetic DNA. Likewise, the 3,000 million letters that make up the entire human genome are enough to code for you – your limbs, your heart, your eyes, your predispositions, albeit with a Shakespearean scope for interpreting meaning. Just as an actor might confer sympathy or antipathy to a character while declaiming the same lines, so too the same gene – the same lines of code – can have very different effects depending on the context. Genetic determinism has little meaning.

Principle is one thing; working out the details took half a century and counting. The first attempts to decipher the code of life were made by physicists, including Crick himself, who sought (and found) a mathematical beauty; but all of them turned out to be utterly wrong. The reality is far messier. The genetic code is riddled with redundancy. An amino acid is encoded by a triplet of letters, known as a codon. There can be anywhere between one and six different codons encoding the same amino acid. All this redundancy seems to limit the consequences of mutations, and so has some biological value. But there was no hiding the bewilderment of the protagonists: the beautiful pared-down symbolism of the double helix vanished without trace in the endless stretches of code that seemed to be bereft of any meaning at all, what became known as 'junk DNA'. Barely 2 per cent of the human genome codes for proteins. Arguments persist over what proportion serves a regulatory purpose; up to 20 per cent might be a generous estimate.[2] The rest seems to have little information content. But whatever the answer, for all its sprawling incoherence, the code is a source of endless fascination. Genomes are a fantastical jungle of digital patterns, mingling sense and nonsense in a similar way to computer code or the internet – meaning riddled with viruses and gobbledegook. Biology has become the study of information with all its quirky content.

That's not intended to be critical. The information revolution has transformed everything in biology, from the study of individual cancer cell lines to the development of embryos, to the deepest reaches of evolutionary time, right back to the first stirrings of life on this planet. Sequencing even overturns cherished ideas of behaviour – sparrows for example, turn out to be far less faithful to their partners than their bearing suggests; nearly a third of offspring are fathered by philandering cheats. Nothing in biology has escaped digital scrutiny. But while our exhaustive exploration of the digital jungle has changed the way we think about everything, it has also helped us forget some lessons from the past. The 'dynamic side' of biochemistry now rarely escapes the pages of dusty textbook histories. It does not seem to add much to the power of information. That's a fallacy. This book aims to show that the flow of energy and matter through cells structures biological information rather

---

2 There are all kinds of interactions between 'regulatory RNAs' that do not involve any proteins at all. RNA is a working copy of a short section of DNA, where the exact sequence of letters in the DNA is transcribed into the same sequence of letters in the RNA copy. This copy can be read off into a protein, or alternatively can bind to other RNA molecules directly. In this case the RNA sequence does not really 'code' for anything; the letters in one string of RNA simply interact with those in another. I'm using the term 'code' very loosely here, to mean the DNA sequence that somehow gives rise to an organism. In any case, in stricter usage, the genetic code should be termed the genetic cipher. Cryptographers can get very cross about this.

than the other way around. Information is obviously important, but it's only part of what makes us alive.

Molecular machines

Biochemistry followed a rather different path, although until recently it too turned its back on the dynamic side. Genes code for the sequence of amino acids in proteins, typically several hundred of them joined together in a string. Yet genes only code for the sequence itself, not the way that this string coils into knots and helices and sheets to give a protein its three-dimensional shape. We have still not unravelled all the rules that govern how a protein folds reproducibly into a particular shape, specified only indirectly by its DNA sequence. Ironically, artificial intelligence algorithms have recently made some progress on this question, but we're not quite sure how they did it. Biochemists, though, have become very good at resolving the architecture of giant protein molecules.

The greatest advances have drawn on the same recondite technique that Rosalind Franklin brought to bear on DNA in the early 1950s, X-ray crystallography, albeit with an enormous increase in power and resolution since then. Perhaps the consummating achievement of crystallography was Venki Ramakrishnan's structure of the ribosome – the astonishing molecular machines, virtually whole factories, that process the genetic code to build new proteins. This is no repetitive structure like DNA, but an enormous assemblage of a quarter of a million atoms, each with its own precisely defined positions. I use the plural here, because proteins are genuinely machines, with moving parts that carry out specific functions. A single protein typically has several conformational states and will switch from one to another at extraordinary speeds – hundreds or thousands of times per second. Understanding precisely how these molecular machines work has preoccupied some of the best minds in biochemistry for decades. Those of us who don't solve the structure of proteins for a living look at the pages of journals such as *Nature* with a mixture of jealousy and admiration, for every issue seems to contain at least two papers detailing the atomic structure of the latest machines. In comparison, even the whole genome sequence of another domesticated species, once all the rage, has begun to pale.

These two themes, information and structure, have combined as the dominant paradigm of medical research in recent decades. We can sequence genomes and search for small differences in sequence between people prone to some condition and those who are resistant. There might be thousands, if not millions, of variations in single DNA letters that predispose to particular diseases, but only a few of these are incriminated so regularly that they stand out as medical targets. If these genes code for a protein, then its structure can be solved in both the normal and defective forms, and it becomes a rational drug target, or may be fixed by gene editing. The idea sounds perfectly reasonable, equivalent to fixing a broken part in a car. But as we've seen, cells are more like cities than cars, and the reality of targeting specific genes is often diabolically complex.

Remember that many proteins are catalysts – enzymes – which convert one molecule into another slightly different form. Vivid as it may be, our earlier metaphor of metabolic flux as traffic transforming in size and shape as it passes down a street fails to emphasise that this doesn't happen spontaneously, at least not in biochemistry as we know it today. Each transformation is catalysed by an enzyme. To force the metaphor a bit too far, we have to imagine the street lined with giant machines (we'd better say they're invisible) that convert one car into another. This succession of giant machines each acts in turn down the street (a metabolic pathway) to change one vehicle into the next. In cells, the output is a product that is useful to the cell in some way – perhaps an amino acid that is used to build a new protein. The problem is that there is often more than one way to do the same thing. Just as there are multiple routes from A to B across the city – different streets we could take to the same destination – so too in the cell there are other ways to get to the same place. In effect, there's more than one street that turns out tractors; around the corner, there's a street that converts jeeps into tractors. If one street gets blocked, then the traffic flow shifts and compensates. But a cell is more sophisticated than a city. If a road becomes blocked, the traffic signals its impotent frustration with blaring horns. Something similar happens in the cell but here the signal of frustration travels straight to the genes, where contingency plans are immediately implemented. Alternative route too narrow to accommodate extra traffic flow? No problem: widen the road. Unlike a city, where harassed local authorities may procrastinate for months, the genes encoding alternative pathways are upregulated in hours, in effect widening the bypasses to accommodate extra traffic.

So the flux of energy and materials through the cell shifts. It might not be quite as good as the main highway, but you could hardly tell the difference. Now think about alternative routes in terms of how a drug acts. Target a particular gene or protein, and the cell redirects metabolic flux to minimise obstructions to function. Just how well it can do this depends on scores of other normally trivial genetic differences, or diet, or other forms of stress such as smoking, weight or age, and these differences account for many of the unpredictable responses to drug treatments. Cancer is a prime example. Drugs typically work for some people but not everyone, or work for a while and then start to falter. The problem is not so much the target itself but the context. Cancer cells grow and evolve. To do that they need to make all the building blocks for growth, which requires constant metabolic flux. Blocking flux at one particular place is like closing a road – it's only a matter of time before the traffic finds another way through, perhaps facilitated by a new genetic mutation, and the cell reverts to uncontrolled growth. Whereas normal cells face constraints in their flux – they have certain tasks in a tissue, such as making hormones or neuro-transmitters, or detoxifying poisons – that's not true of cancer, as we'll see. Cancer cells simply switch to another flux pattern and keep on growing. The problem is not information alone; the deeper, underlying problem is flux. It's the dynamic side of biochemistry again.

Satnav metabolism

There's a new name for this that reflects the modern 'omic' age: metabolomics. We have known all the steps in the major meta-bolic pathways for decades – they were laboriously worked out, step by step, from the 1930s onwards, with a leap forward in the post-war years, when radioactive tracers enabled the fate of specific carbon atoms to be tracked (as we'll see in Chapter 2). Metabolomics is much the same thing, but now with the aid of powerful techniques such as mass spectrometry. Instead of seeking the commonalities – the same metabolic pathways in different cells – metabolomics looks for the differences: how does the flux of materials and energy through one of my heart cells differ from that in one of yours? It is closer to a satnav-enabled road map, showing live congestion, but it's still a snapshot in time. We can take a sample of cells and look at the distribution of flux at one moment, or over a short period of minutes or hours.[3] Will it be the same next week, next month, next year? Start over. We are a long way from rising above the cell and picturing the combined flux of energy and materials in real time over a lifetime. Metabolism remains invisible and elusive, even if we know it keeps the cell alive.

But these subtle differences in flux also emphasise one dramatic way in which cells differ from cities. Different cities are similar in that they all have roads, but the actual road map obviously differs from town to town. That's partly true of cells too, but perhaps the most extraordinary fact is that all cells share the same basic road plan, at least for the city centre itself. What differs is the congestion or the size of the roads, not the layout. The unity of biochemistry means we all share the same city-centre map. What defines this map? You might say 'genes', but that is far from true. We'll see that genes did not 'invent' metabolism, but the reverse. In any case, genes change and evolve, but the pathways they are supposed to code for remain essentially unchanged. As the poet Edna St Vincent Millay wrote, 'life isn't one damn thing after another, it's the same damn thing again and again'. A bacterial cell living deep down in the crust of the earth makes its letters for DNA through the same succession of steps that you do, even if many of its genes have diverged almost beyond recognition. Genes are far more malleable than metabolism. Likewise, if a gene in a cancer cell

---

3 Even this can be hard to interpret. A high concentration of some intermediate in the cell does not necessarily imply high flux – it might reflect a high rate of flux, with the intermediate being constantly replenished at high levels, or it could reflect the opposite – almost no downstream flux, so the intermediate gradually builds up; in effect, a car park. To understand which one requires a lot of context and subtle interpretation. There are times when it feels as if metabolomics should just be called gnomics.

mutates, and now promotes uncontrolled cell growth, this does not open up some new metabolic pathway, but rather diverts flux down existing pathways, even if it dramatically reverses the flow down a one-way street. Traffic flow might change but the map itself rarely does.

The reason that city street-plans differ but the central metabolic maps of cells do not is quite simple: cells descend from a common ancestor, but cities do not. Cities are similar by analogy, not homology. When we think about inheritance we tend to think about genes, but to leave any descendants a cell must be capable of growing, repairing and ultimately replicating itself, and to do that it needs a fully functional metabolic network. To be alive means to have a continuous flow of energy and materials through this whole network, nanosecond by nanosecond, minute by minute, generation after generation. We do not merely inherit inert information in the form of genes – our inheritance includes this living metabolic network in the egg cell, a flame passed from generation to generation, without pause, right back to the emergence of life. Core metabolism has changed little in part because it was never powered down in its four-billion-year history. The genes are custodians of this flame, but without the flame life is – dead.

Yet despite its unceasing flow, metabolism has been turned on its head. There is no better example of the crazy contingency of evolution, or the ability of life to cobble together a workable solution to the utter transformation of conditions on Earth, from the suffocating anoxia of the first two billion years to the energised atmosphere of the age of animals, without which we could not exist. At the heart of the cell is a merry-go-round of energy and matter known as the Krebs cycle, after the venerated biochemist Sir Hans Krebs, who first conceived this iconic cycle of reactions in the 1930s. It's also sometimes called the 'citric acid cycle' or the 'tricarboxylic acid cycle', but let's stick with the more personable name in this book. Unlike most metabolic pathways, which run in straight lines, the Krebs cycle leaps out from our metabolic maps with a Platonic sense of perfection, the perfect circle at the centre of everything, yet elusive still in meaning more than eighty years after its discovery. Elusive, in part, because biochemistry moved on long ago, enchanted by the marvellous mechanics of molecular machines. But elusive most of all because this apparently perfect cycle of energy and matter in fact conceals a strained balance of opposites, a yin and yang, which touches on all aspects of life.

The Krebs cycle

Generations of biochemistry and medical students have been obliged to learn the steps of the Krebs cycle by rote. Despite its iconic status it has earned little love or real understanding. That's partly the problem with visualising biochemistry: this is an invisible and abstruse set of reactions, each step an ostensibly trivial rearranging of carbon, hydrogen and oxygen atoms (see pages 286–7). But beyond that, even its true function is obscure. The textbooks tell us that the Krebs cycle generates energy by stripping out hydrogen atoms from the carbon skeletons of food and feeding them to the ravenous beast that is oxygen (well, they might not put it in exactly those terms). This is the process of cellular respiration. The energy released at each step is ingeniously captured and put to use in the cell, while the inert carcasses of water and carbon dioxide are discharged into the outside world. But why is it a cycle at all? Why not just a few simple steps? Krebs himself suggested one plausible answer: burning very small carbon skeletons is not an efficient process, he argued, so a cycle was necessary. The later discovery of bacteria that burn tiny carbon skeletons perfectly well disproved that idea.

The *raison d'être* for the Krebs cycle is muddied even further by the fact that the cycle supplies many of the basic building blocks for the fabric of the cell. Most amino acids are made directly or indirectly from molecules in the Krebs cycle. So are the long-chain lipid molecules needed to make cell membranes. New sugars are made from the Krebs cycle too. Even the 'letters' of DNA (termed nucleotides) are made from sugars and amino acids, and so also derive from the Krebs cycle. I could go on but suffice to say that the Krebs cycle is the engine of biosynthesis, driving cellular growth and renewal. But why use the same pathway to create and destroy, to burn and renew? Even a phoenix can't do both simultaneously. Most metabolic pathways separate biosynthesis from breakdown for the simple reason that flux can't go both ways at once. Yet the same molecule in the Krebs cycle can be converted into an amino acid and used to build a protein, or shredded and burnt in the furnace of respiration, to generate cellular energy. Is there any rhyme or reason to this conflicted merry-go-round of energy and matter?

A decade ago the answer to this question would have seemed too esoteric for most researchers to worry about. Asking 'why?' has historically been dismissed as rank speculation by biochemists. But then it turned out that molecules accumulating in the Krebs cycle can signal the state of the cell to the genes, switching on or off hundreds or even thousands of genes. Far from being dusty textbook biochemistry, we now know that different patterns of metabolic flux through the Krebs cycle can generate powerful, if ambiguous, signals. The same signals that help diving animals such as sea turtles survive underwater for hours without oxygen can also promote growth and inflammation in cancer. Mutations in some genes linked to the Krebs cycle crop up again and again in aggressive tumours. Flux through the Krebs cycle is also linked with our likelihood of surviving a heart attack, for example if we have diabetes. Perhaps that should not be surprising. Anything that disturbs the availability of oxygen affects not just respiration but also flux through the Krebs cycle. Breathing is an immediate life-and-death problem, but so too is the turnover of the molecules that make up our bodies. The problem in all these disparate cases is how to balance the yin and yang of the Krebs cycle – how to offset the needs of energy generation against the synthesis of new organic molecules. This question brings the Krebs cycle into sharp modern focus but does not explain why the cycle has a yin and yang at all.

In my view, we can't hope to understand what goes wrong in cancer or Alzheimer's disease if we do not know why the Krebs cycle is so central to the flow of both energy and matter in cells. What rules govern this flux? Genes are only part of the picture. Let's put aside our traffic metaphor and think about the flow of water in a river. If genes channel metabolic flux like riverbanks channel water, then the genes do not determine the trajectory of flow any more than riverbanks determine how water cascades down from the mountains to the oceans. That is determined by the properties of water, the power of the sun and the landscape – by evaporation from the oceans, rainfall high up in the mountains, the softness of the underlying rocks, the loose, rolling electrical bonding between the water molecules that make it a rushing coherent liquid, the incessant tug of gravity. We might channel a river's flow between raised banks in our cities, but even the cleverest construction is of little use against a serious flood. Metabolic flux is equivalent. Genes code for protein catalysts, but catalysts are not magic – they simply speed up reactions that happen spontaneously.

The products are the products of chemistry, which will form anyway in the absence of a catalyst, albeit at a tiny fraction of the speed. The driving force for metabolism is thermodynamics. This is an intimidating term, but in this context only means the chemical need to react (to dissipate energy) in the same way that water needs to flow downhill.

If the Krebs cycle is ordained by thermodynamics, then it should take place spontaneously in some suitably propitious environment, even in the absence of genes. That idea was once dismissed as 'if pigs could fly' chemistry, yet revolutionary new work shows that at least some of the cycle can indeed just happen, catalysed by rocks and metals, rather than by proteins encoded by genes. These new discoveries have renewed interest in the actual chemistry that Krebs first laid out decades ago, but in a primordial setting he could scarcely have imagined, and backwards. The inner logic of metabolism is beginning to take shape. Much of it is imposed by thermodynamics; some is facilitated by catalysts. Some is refined by genes. And part stems from the vicissitudes of life itself, which forced evolution down improbable paths, while transforming our geologically restless globe from a sterile, anoxic planet into the living, high-octane world of today.

This is the story of our planet from within our cells – what makes us the way we are, and ultimately why we age and die, each of us in our own tantalisingly different ways. I appreciate that biochemistry is for many people an obscure subject, full of runic symbols that suggest a priesthood intent on concealing the path to meaning. Yet nothing could be further from the truth. What is more meaningful than the chemistry that brings life alive? Or the chemistry that sends us to our death? Or the chemistry that forges our conscious self? Yet astonishingly, all of this is the same chemistry! I will try to bring this living chemistry to life in plain terms, but without glossing over the important details – we will take a journey to the frontiers of human understanding. I admit our voyage will not always be plain sailing, but I hope that you will find the rewards worth it. This voyage is my own attempt, as a biochemist, to find the inner meaning of the Krebs cycle: why it is still spinning at the very heart of life and death today. Please join me.

Setting sail

We will start by getting acquainted with the Krebs cycle itself – the basic chemistry you will find in any biochemistry textbook. I will enliven the dry dance of molecules with a tribute to some of the pioneers of biochemistry, men and women who imagined such ingenious, creative ways to read the inscrutable matter of life that they should be household names in any culture where science holds its head high alongside the arts and humanities. Even today, I am using the modern equivalents of their methods in my own lab. But there's another reason why I am telling you their story: these are the exalted pioneers of biochemistry, and even if they are little known to the general public, they cast long shadows across their field. Not all their ideas were correct. Scientists are human. We are only now seeing beyond their interpretations because we, too, are only human. Too many of their views became dogma. The Krebs cycle is at once the same and far more than Hans Krebs ever knew.

In Chapter 2, we will begin to deconstruct the textbook view. We will see how dogmas can set back fields by decades, and that we are by no means immune to them today. Still you will read that the Krebs cycle is about the oxidation of food-stuffs, and especially glucose. You will read that photosynthesis makes glucose, while respiration burns it by way of the Krebs cycle. Sugar, sugar, everywhere. This view is badly miscon-strued, for it places the heart of metabolism, the heart of life, in the sugary chemistry of photosynthesis and respiration. On the contrary, the core of metabolism lies in the Krebs cycle – as practised by ancient bacteria that run the cycle backwards (or forwards, from their point of view) reacting the simple gases hydrogen and carbon dioxide to generate the universal precur-sors of life.

In Chapter 3, we'll go right back to the beginning, to see how this same chemistry can take place spontaneously in nature, and specifically in deep-sea hydrothermal vents, where the reaction between H_2 and CO_2 is facilitated by steep gradi-ents of protons across thin inorganic barriers, with structures similar to cells. We'll see that this chemistry can in principle – and recent experimental work shows in practice – drive almost all the core metabolism of cells, giving rise to the building blocks of genes (the 'letters' of DNA). I shall argue that genes emerged in protocells capable of a rudimentary form of hered-ity, giving meaning to the emergence of genetic information from the beginning. We'll touch on recent work from my own research group, who are beginning to make exciting progress on the origins of genetic heredity in this setting.

Chapter 4 will ask how this spontaneous chemistry at the origin of life came to be the closed cycle that we know today. The idea of a universal cycle is misleading: even in humans, the cycle is often more of a roundabout, with flow entering and exiting everywhere, even coursing in opposite directions through different bits of the cycle. Tight energetic constraints on the early Earth forced metabolic efficiency and cooperation on bacteria, until the evolution of photosynthesis changed life forever. Yet the link between rising oxygen levels – the ener-gising waste product of photosynthesis – and the geologically abrupt emergence of animals in the Cambrian, around two billion years later, was dislocated by some of the most diabolical global conditions in the history of our planet. We'll see that the need for metabolic efficiency might have fostered cooperation not only in bacteria, but even between the mutually dependent tissues of early animals creeping through sulfurous sludges, their lives teetering on the brink.

Not much is known about how metabolic flux is balanced between different tissues, but we know much more about what goes wrong when cells revert to selfish behaviour in cancer. In Chapter 5, we'll see that the standard view of cancer as a disease of the genome driven by genetic mutations is not the whole picture, and indeed verges on being a latter-day dogma. Mutations linked with cancer are commonly found in normal tissues, while cancer cells often cease dividing if placed in normal tissue. In fact, age is the biggest risk factor for cancer. We'll see that cancer emerges from the yin and yang of the Krebs cycle as we get older – the need to use the same pathway for biosynthesis as well as energy generation. As the flux through our Krebs cycle becomes more sluggish with age, intermediates such as succinate accumulate and trigger the ancestral pathways for dealing with low oxygen levels, driving inflammation, cellular growth and proliferation, all of which promote cancer.

In Chapter 6, we'll examine why Krebs-cycle flux gets more sluggish with age, unmasking different age-related diseases in each one of us. Much of the answer can be ascribed to a gradual depression of cell respiration with age, which varies with our lifestyle (such as diet or exercise), as well as how effectively our two genomes work together – the genes in the mitochondria alongside those in the nucleus. I'll give a new take on the free-radical theory of ageing from my own research, which might explain why birds live so much longer than equivalent mammals, and why antioxidants are not helpful. And we'll see why full brain function needs the full Krebs cycle, and conversely why failing cellular respiration is linked with conditions such as Alzheimer's disease.

Finally, the Epilogue will touch on the hardest problem of all, the stream of consciousness. By now, it should come as no surprise that the moment-by-moment flux of metabolism that makes us alive should be interwoven with our innermost feelings of self, perhaps going right back to the first stirrings of life. How exactly? We'd better begin at the beginning.

1

DISCOVERING THE NANOCOSM

Burlington House, Piccadilly, 1932. Its stately Victorian façades are glittering with light at the fag-end of a particularly dismal November. A spry, silver-haired gentleman, immaculately dressed and moustachioed, is coming to the end of his anniversary address as President of the Royal Society of London. Just three years after his Nobel Prize for the discovery of vitamins, Sir Frederick Gowland Hopkins is now a sprightly 71, his intellectual vigour undiminished. He had begun his address with a celebration of nuclear physics. Earlier that year Chadwick had proved the existence of the neutron, and Cockcroft and Walton had split the nucleus, releasing the power of atomic energy. Kudos to the Cavendish Laboratory in Cambridge – truly an *annus mirabilis*. The phenomenon of transmutation seemed to be at hand in full reality, Hopkins had remarked, relishing his words. After all, the lithium nucleus had been converted into two helium nuclei – nuclear fission. Who knew then what the next decades would bring.

With a degree of trepidation that he's careful to offset by his choice of words, Hopkins moves on to the emission of radiation by cells, which he notes might even accelerate cell division. Does he hear murmuring in the audience? No matter. Science must push the boundaries. He is eager to close with the most brilliant work from his own field of biochemistry. Science is international. He picks out the young Krebs of Freiburg-im-Breisgau, whose revelatory work has just the degree of unexpectedness to be expected of biochemical phenomena (he allows himself a flicker of pride in his wording here) – work that shows why experimental biochemistry must remain an independent scientific discipline. The audience knows he's an unabashed advocate for his fledgling field. He is content to live up to it. Hans Krebs had shown how biochemistry can escape from the dogma that chemical methods cannot be used to study life because they instantaneously convert the living into the dead. By adding precise quantities of amino acids to thin slices of living tissues and making meticulous measurements of the gases emanating, Krebs had shown how cells make the waste-product urea, excreted in urine, through a cycle of reactions. This was not dead chemistry, but a dynamic living process. In the audience, Hopkins's many friends are smiling. They know his obsession with the dynamic side of biochemistry. And they also know he's right.

There is no more beloved figure in science than Sir Frederick. Nobel Prize, Knight of the Realm and soon to be elevated to Order of Merit: a pillar of the establishment. But he always felt an outsider. He had been expelled for truancy (or technically 'his removal was advised') from the City of London School. He had simply failed to turn up one morning, for reasons he could never explain to himself, and then, shrinking from the inevitable shame and punishment, persisted in his truancy for weeks, hanging out in shipyards and museums and libraries. He'd ended up in a new school that had barely tested him at all, before taking an unrewarding (and indeed unpaid) apprenticeship in an analytical chemistry laboratory. But it turned out he was a brilliant and meticulous chemist. He took several short courses

at which he excelled before finally getting a break – a dream
job in forensic medicine at Guy's Hospital in London. That led
him to Cambridge at the end of the nineteenth century, where
in due course he was elected its first professor of biochemistry.
It had been an unusually circuitous route, which omitted the
orthodox training in Germany, where the founding fathers of
biochemistry had plied their craft. No doubt that only added
to his sense of being an outsider. It must have fostered his origi-
nality too. He had nothing of their authoritarian approach to
science but rather earned the love of those around him.

Hopkins nurtured a joyously free, productive and non-hier-
archical lab environment. He had been backed consistently by
the nascent Medical Research Council (MRC), but they were
perpetually frustrated by his stubborn refusal to dedicate his
time and efforts to applied medical research, neither taking
his own early work on vitamins forward, nor expecting others
to address practical problems. Instead he collected a group of
incalcitrants, who seemed to work on whatever they pleased,
including the dazzling polymath J. B. S. Haldane, who was
temporarily dismissed from Cambridge for gross immoral-
ity (cohabiting with a married woman), and several brilliant
if equally unmanageable women, notably Marjory Stephenson and Dorothy Needham. Known affectionately as Hoppy
to his colleagues (which I doubt impressed the MRC either),
Hopkins did occasionally influence their direction of research,
for example convincing the young Marjory Stephenson to
work on bacterial metabolism. Her pioneering studies (most
famously on bacteria living in stagnant river waters, causing
more anguish at the MRC) forced open the doors of the Royal
Society; she was elected its first female Fellow in 1945, alongside the crystallographer Kathleen Lonsdale.

But Hopkins's real mission was the development of

biochemistry as an experimental discipline, with its own methodology and way of seeing the world. It was vibrant and fun. The lab's journal, *Brighter Biochemistry*, included compilations of verse (Haldane wrote an annual report in rhyming couplets), exam questions from the future, cartoons and cautionary tales, such as laments for 'Jane who had no bacteriological technique and so perished miserably' and 'Belinda who broke everything and left the laboratory under lamentable circumstances.' Don't be fooled by their irreverence. These were serious minds at play, and Hoppy's laboratory nurtured some of the most imaginative and original scientists of their generation, including a number who went on to win Nobel prizes.

Hans on board

This was the world into which the young Hans Krebs arrived in 1933. He had not intended to leave Freiburg, of course, but had been relieved of his position, like almost all those of Jewish ancestry, after Hitler became Chancellor in January 1933. Krebs soon realised he needed to get as far from Germany as possible; even the offer of a position in Zurich seemed a little too risky. Rarely has the attainment of fame (and in particular the recognition from Hopkins in his anniversary address to the Royal Society) been more fortuitously timed. Hopkins was one of a group of the great and good – forty-one eminent intellectuals, including H. A. L. Fisher, Margery Fry, J. S. Haldane (the equally distinguished father of J. B. S.), A. E. Housman, John Maynard Keynes, Ernest Rutherford and J. J. Thomson – who in May that year had founded the Academic Assistance Council to support scholars who had been obliged to relinquish their posts in Germany. The founders defined their goal as 'the relief of suffering and the defence of learning and science'. Krebs was the first to be backed by this scheme, arriving in Cambridge in June with a cargo of thirty manometers, his equipment for measuring minute changes in gas pressure.

You can imagine that Hopkins's lab was very different to anything that Krebs had known in Germany. Years later, Krebs reflected on what he called 'the British way of life', writing that 'The Cambridge laboratory included people of many dispositions, convictions and abilities. I saw them argue without quarrelling, quarrel without suspecting, suspect without abusing, criticize without vilifying or ridiculing, and praise without flattering.' I hope that Britain today does not forget those values.[1] Krebs felt immediately at home (indeed he was literally welcomed into scientists' homes) and applied himself to mastering the English idiom. There's a charming story that this earnest young man burst into the lab one afternoon, as giddy as a child, and declared that he had been 'pulling Marjory Stephenson's legs'.

But for Krebs, all this was merely a backdrop to the paramount importance of his work. He was beginning to address

---

1 For balance it should be said that there were also some protests outside the lab against foreigners taking scarce British jobs during the Great Depression. The parallel with modern times is uncomfortably close. But here is a more edifying modern parallel. Lord John Krebs, the distinguished son of Hans, tells a beautiful story to mirror his father's. After reading a newspaper article about the high price that Nobel medals can fetch at auction, he felt obliged to either donate his father's medal to a museum, or seal it in a vault, or sell it and put the proceeds to a good cause. He chose the latter, and put the funds into the Sir Hans Krebs Trust, to support today's refugee scientists forced to flee from their home country. In collaboration with CARA (Council for At-Risk Academics, the descendant of the Academic Assistance Council) the Sir Hans Krebs Trust has since 2015 supported a number of scientists fleeing from Syria and other countries. You can find more information here: https://www.cara. ngo/sir-hans-krebs-trust-cara-fellowships/. Should you wish to donate, please alert Stephen Wordsworth that the donation is for the Sir Hans Krebs Trust (email: Wordsworth@cara.ngo)

the most pressing question in biology at the time (and argu-ably still a vital question today) – how does respiration work? The broad idea is simple enough: respiration burns food in oxygen to generate the energy we need to live. That much had been known since Lavoisier, before the French Revolution, who showed that respiration and combustion are exactly equivalent processes (*exactly*; he weighed everything with obsessive preci-sion). Respiration and combustion both involve the complete oxidation of organic matter by oxygen to release energy, ulti-mately as heat. But for living organisms, not all the energy is released as heat immediately; some of it is first captured and used to drive work (technically, this is 'free energy'), before it, too, is eventually dissipated as heat. That's why animals hasten the heat death of the universe. In the case of sugars such as glucose, the overall reaction for both combustion and respira-tion can be given as:

$$C_6H_{12}O_6 + 6\,O_2 \longrightarrow 6\,CO_2 + 6\,H_2O + \text{energy}$$

$$\underset{\text{glucose}}{} \quad \underset{\text{oxygen}}{} \qquad\qquad \underset{\substack{\text{carbon} \\ \text{dioxide}}}{} \quad \underset{\text{water}}{} \quad \underset{\text{heat}}{}$$

What was not known was how on earth the energy released by this reaction was captured and put to work within the cell, to drive everything that life does, from moving around to thinking. Plainly, the energy had to be released in small chunks or we'd simply go up in flames, but the way in which this happened was unknown; even the location within the cell had not been established when Krebs started thinking about the question. We now know that respiration takes place in the mitochon-dria, the so-called 'powerhouses' of the cell; but every scrap of knowledge along the way was hard won. And knowledge comes tinged with the stories of how it was won. That's only human.

The harder the discoveries, the better the stories – but also the more scope they have to colour our thinking. Krebs started out lacking any blueprint for how respiration might work, and he ended up with a detailed set of steps in his famous, justly eponymous cycle. Yet his approach also coloured the way successive generations have thought about respiration ever since, to the point that the Krebs cycle has become virtually synonymous with respiration. As we saw in the Introduction, it's much more than that; the Krebs cycle is not only about respiration but also, and equally, about fashioning the building blocks of life. An overly narrow view of the Krebs cycle has obscured our understanding of the origin of life, cancer and even consciousness, as we'll see in this book.

Slices and gases

Krebs's experimental approach derived from that of his mentor, the great German biochemist Otto Warburg (whom we'll meet again in Chapter 5, on cancer). Meticulous and brilliant, Warburg was also domineering and authoritarian; he gave his tight-knit group little freedom to develop their own ideas and expected extreme punctuality for six days a week, with Sundays used for writing up results and preparing the following weeks' experiments. Krebs had flourished, working with Warburg for four years in what amounted to an apprenticeship. He maintained similar working habits throughout his life, but more importantly, he developed Warburg's ingenious methods.

The key method that Warburg had pioneered was to measure the escape of gases from thin slices of tissue, cut using a razor blade. When practised skilfully, the slice was thin enough for oxygen to penetrate fully by diffusion, while the cells in the centre of the slice were undamaged, functioning more or less normally. The slices were then placed in a solution with a similar composition to blood plasma, in a glass flask joined to a manometer (which measures gas pressure like a barometer) and sealed tight. Under these conditions, gases could bubble out of the tissue slice, which would increase the pressure inside the manometer and alter the height of liquid in a U-tube. Of course, accurate measurements required careful calibration. For example, because the pressure measured depends on temperature, the apparatus had to be submerged in a water-bath controlled by a thermostat. The apparatus needed to measure one biochemical reaction filled half a room, and still gave an indirect view of what was actually happening, for just a single reaction step. It took unimaginable patience and skill to figure out exactly what was going on in the multiple steps of a biochemical pathway.

This method gives a good feel for how indirect and convoluted it is to measure anything in biochemistry. We can't 'see' any of these small molecules even today (where more sophisticated methods are still harder to interpret). Yet this was how biochemistry came to be understood; the bottom line is that it worked. The rate at which various gases bubbled out from freshly cut tissue slices could be monitored precisely over time, giving unprecedented insights into living processes. Just think about how cells respire. As you can see in the equation on page 32, respiration consumes oxygen and releases carbon dioxide, so the rate of respiration can be gauged by measuring the change in gas pressure over time.[2]

---

2 If you are sharp-eyed you might notice that the rates of oxygen consumption and CO_2 release counterbalance each other exactly, so there is no overall change in pressure in this case. But there's a clever trick. The CO_2 can be absorbed by a solution of potassium hydroxide, so that it exerts no gas pressure. An equivalent method is used today industrially as a CO_2 'scrubber' for carbon capture;

The experiment for which Warburg won the Nobel Prize in 1931 was just so beautiful I have to tell you about it. Warburg knew he could block respiration by using the gas carbon monoxide (CO), which in cells binds to metal atoms in the catalyst that converts oxygen to water. The binding of CO therefore halts the final step of respiration. On a manometer, the binding of CO registers as zero change in gas pressure, because oxygen is no longer consumed. Here's the clever bit. CO can be detached from the catalyst through the absorption of light. That kick-starts respiration and oxygen consumption over again. But not all wavelengths of light work equally well; some are absorbed by the catalyst, which then releases its bound CO, while other wavelengths just bounce off, leaving the CO bound. Warburg used light sources that emitted photons with specific wavelengths, such as mercury or sodium vapour lamps, or burning magnesium salts. By measuring the rate of respiration at different wavelengths of light, he could reconstitute the absorption spectrum for the catalyst (what he called the ferment), showing that it was a haem pigment, closely related to haemoglobin, the pigment that transports oxygen in our own red blood cells, and not dissimilar to chlorophyll, the pigment used in photosynthesis.[3] In fact, Warburg went even further than that, showing that

_____

the CO_2 is converted into carbonate, which is eventually precipitated out as calcium carbonate (in effect, limescale). For manometry, instead of measuring the change in pressure exerted by both CO_2 and oxygen, the manometer only measures the diminishing pressure exerted by the consumption of oxygen.

3 Carbon monoxide poisoning can kill you by binding to the iron in the haemoglobin of your red blood cells and displacing oxygen, but if you survived that, CO would stifle respiration in your cells for the same reasons. The haem protein that Warburg discovered is now called cytochrome oxidase, 'cytochrome' meaning 'cell colour'. Without cytochromes, cells are nearly colourless (in fact they're a pale yellow from flavins, also discovered by Warburg). At much lower levels, CO is an important cellular signal that operates in part by modulating the rate of respiration, as does another important gaseous signal,

spectra similar to both haemoglobin and chlorophyll could be generated by simple chemical changes to his ferment. From that, remarkably, he concluded that respiration arose before photosynthesis: 'blood pigment and leaf pigments have both arisen from the ferment – blood pigments by reduction, leaf pigment by oxidation. For evidently, the ferment existed earlier than haemoglobin and chlorophyll.' He wasn't so far from the truth, as we'll see in Chapter 4.

I've never forgotten reading Warburg's Nobel address, where he described all this, while I was sitting outside my tent near Cader Idris in Wales one beautiful morning, unable to tear myself away towards the beckoning mountain. My head was singing along with the birds in the hedgerows around. It was just so clever. This is experiment as work of art, creative imagination allied to virtuosic technique, giving a beautiful insight into the workings of nature. This is why some scientists are put on pedestals, even though, like great artists, they often have all the vices of their virtues. That was true of Warburg more than most.

Yet even this experiment-as-work-of-art was far from elucidating what actually goes on in respiration. Really, all that it said was that one of the catalysts involved (Warburg characteristically asserted it was the only one of relevance) contained a porphyrin pigment similar to haem. But what about all the other steps going from glucose to oxygen? How many steps were there? What were the intermediates? You can see in my equation above that one glucose molecule, with its six carbon atoms, is converted into six molecules of carbon dioxide and

---

nitric oxide (NO). These gases can control the rate of respiration so exquisitely that they alter the local concentration of oxygen. This trick is employed wonderfully by fireflies to generate flashes of light using the oxygen-dependent enzyme luciferase.

six molecules of water. Was each CO_2 yanked off individually, so that the six-carbon chain of glucose became progressively shorter as each CO_2 unit was released? What about the water? Presumably, the twelve hydrogen atoms in glucose were all extracted, and ultimately transferred onto oxygen to form the water (H_2O). This is the source of almost all the energy from respiration – the hydrogen is burned in oxygen. That's how rockets are powered. Plainly we have plenty of power, but that says nothing about how hydrogen is extracted from glucose or its breakdown products, how many steps are involved, what the intermediates are, and how the energy is captured and converted into a usable form. Virtually none of that was known when Krebs began his work. Yet these steps add up to the difference between a conflagration and a living organism. That's what Krebs set out to understand.

To me, the most amazing thing is that Krebs could modify Warburg's methods for measuring gases emanating from tissue slices to piece together most of these missing steps. It was not so much a single breath-taking experiment as a magisterial vision of what was possible using a single method, coupled to a programme of work that would make normal folk sag at the prospect. Krebs sometimes ran as many as ten manometers simultaneously in a single study. He would add known amounts of various substances that could be respired, and measure the rate of respiration in the presence or absence of respiratory inhibitors. The devil really was in the detail. For a modern-day biochemist, it is hard to appreciate just how little was known, how utterly opaque the workings of the cell. As usual in science, some experiments worked well first go and others did precisely the opposite of what had been expected. Sometimes that was because the tissue slice hadn't been washed correctly, or the solution wasn't fresh enough, or some other problem; but occasionally it was because the working hypothesis was wrong, and then suddenly a new avenue opened up, leading, hopefully, to a real advance in understanding. Krebs's lab notes are bristling with double or even treble exclamation marks, as unexpected findings raised his pulse. Even this most measured of men could scarcely contain his excitement. Science is an emotional roller-coaster no matter how much scientists strive to build an objective and unemotional framework. It's a wholly human pursuit. And so it was that day after day, week after week, year after year, Krebs slowly pieced together the detailed steps of how food is burned in respiration.

Deep breathing

The work that had initially made Krebs's name was on the synthesis of urea, a nitrogen-containing compound that is excreted in urine, and which is formed from the breakdown of amino acids, the building blocks of proteins. Ironically, Krebs had been trying to get away from Warburg's focus on respiration; he needed to establish his own research agenda. But his questions kept on cycling back round to the same place. Krebs was wondering what happened to amino acids after they had passed on their nitrogen to form urea. By definition, amino acids contain nitrogen in the form of an amino ($-NH_2$) group.[4] Krebs inferred

---

4 A 'group' refers to a few atoms bound together within a molecule, which tend to stick together in the same arrangement even when transferred to another molecule. Groups are chemically semi-stable configurations that are found in multiple molecules, such as the amino group in all amino acids. While atoms are the basic building blocks of molecules, atoms primarily assemble into groups, which together make up larger molecules. It goes on. Molecules such as amino acids can be linked together to form macromolecules, such as proteins, but the actual chemistry a protein carries out (for example, at the active site of an enzyme) depends on the specific chemical properties of the groups involved.

that this amino group was liberated as the gas ammonia (NH_3) in the kidneys, eventually to be excreted as urea. The carbon 'skeleton' that remains of the amino acid is now termed a carboxylic acid, composed only of carbon, hydrogen and oxygen. Krebs was trying to measure how much ammonia formed in his tissue slices but could only measure the overall change in gas pressure using his manometers. To make meaningful measurements, he needed to take into account the rate of respiration, which as we've seen consumes and releases gases too. And that's where matters became interesting.

Krebs found that adding amino acids increased the rate of respiration in his tissue slices, as did adding carboxylic acids. But adding both of them together did not have an additive effect on the rate of respiration; rather, the breakdown of amino acids (and the release of ammonia) was *inhibited* by adding carboxylic acids. You can imagine how many months of hard work it took Krebs to come to this anodyne conclusion. But what did it mean? Presumably, the excess carboxylic acids simply got in the way of amino-acid breakdown. That made sense, because when amino acids lose their nitrogen, they become carboxylic acids, as we've just noted. In which case, both types of molecule had to be respired via the same pathway, which was now swamped with carboxylic acids. Clearly carboxylic acids are important in respiration, and they are a step downstream from amino acids, a step closer to oxygen in the controlled furnace of respiration.

That caught Krebs's attention because the breakdown of glucose also generates carboxylic acids. The first few steps in the breakdown of glucose had been established by the early

---

Some are more reactive than others. We'll consider the diverse chemical properties of the different groups in a simple molecule in a moment.

1930s. As noted above, a molecule of glucose contains six carbon atoms, with the general formula of $C_6H_{12}O_6$. Let's simplify the terminology here and just call it a C6 sugar, meaning there are six carbons in the skeleton of the molecule. Glucose is split (through a baroque sequence of steps called glycolysis that we can put aside here) into two three-carbon molecules known as pyruvic acid: a C3 carboxylic acid that is also generated by the breakdown of some amino acids. It seemed that everything was being funnelled into the formation of carboxylic acids. Krebs was getting excited. His lab books bristled with more double exclamation marks than ever. Surely, the secret of respiration lay in how these carboxylic acids were burned in oxygen. What were the steps? How was the energy released captured and put to use?

It is time to get acquainted with the molecular protagonists of this book. What exactly is a carboxylic acid? Let me draw one for you. This is the three-carbon pyruvic acid:

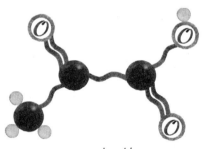

pyruvic acid

I'm depicting carbon atoms as black balls, hydrogen atoms as small grey balls, and oxygen atoms as white balls containing the symbol O. I've drawn the atoms roughly to scale. Oxygen is slightly smaller than carbon, for example, because it has eight rather than six protons in the nucleus, and the extra positive charge draws in the surrounding cloud of electrons a little closer, making the atom smaller. I'll use the same style throughout this book. I hope that it's less intimidating than standard chemical representations for anyone with a bit of a phobia about chemistry. But I'd also like to think of these illustrations as affectionate portraits of the molecular heroes of this book, nay, of life itself, so let's tarry a while and make ourselves better acquainted. When people say that life on Earth is 'carbon based', carboxylic acids are your basic hand of cards.

Portraits capture the character of a person, along with a few of their prominent features; a large nose, squinting eyes, laughter lines and greying beard in my own case. It's harder to read the face of a molecule, but it can be done. Each carbon atom in pyruvic acid has a different character, with its own distinct behaviour and tendency to react. Consider the left-hand carbon atom here for example: it has three hydrogen atoms attached. That's normally quite stable and inert. Stolid. But there's lots of pent-up energy stored in those hydrogen atoms. How easy it is to get at depends on the molecular surroundings. Is this stout-bellied group associated with the molecular equivalent of a pin-striped suit, or a peg-leg and an eye-patch? It's closer to the pirate in this case. That's because the *middle* carbon atom in pyruvic acid has a double bond to oxygen, which likes nothing better than getting into a fight; the extra bond tends to throw its weight about. That's what makes pyruvic acid a touch unstable and reactive. Now the middle oxygen in my picture is staring back at me like Mad-Eye Moody. This bad-ass oxygen is technically called an 'alpha-keto' group, but we don't need to worry about the terminology; we just need to know that it makes things happen. This part of the molecule reacts to form an amino acid, for example, where the oxygen is switched with ammonia to form an amino group, or vice versa, one of the most important transformations in biology (and what Krebs was studying when he was working on the breakdown of amino acids).

What about the third carbon, then, at the right-hand side? That also has a double-bonded oxygen, along with another oxygen bound to a hydrogen, yet this arrangement has a very different character. In fact, it's this group that defines pyruvic acid as a carboxylic acid. An acid is a molecule that wants to cast off a proton whenever it can – a proton being the charged nucleus of a hydrogen atom. Here's what happens when you put pyruvic acid in water:

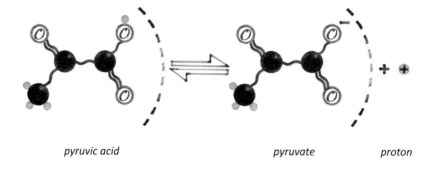

pyruvic acid   *pyruvate*   *proton*

Note that the relative size of a proton is much smaller than depicted here, but if I drew it to scale it would be smaller than a microdot. I've drawn it the same size as a hydrogen atom for clarity.

In this case, I've drawn a dotted line next to the carboxylic acid group, which consists of a CO_2 (one carbon and two oxygen atoms) plus a hydrogen atom attached to one of the oxygens. All carboxylic acids have at least one group like this by definition; there can be two or three of them. Now when a carboxylic acid loses a positively charged proton, it's left with a negative charge (–) which I've shown on the top oxygen for clarity. But in reality this charge is smeared in space between the two oxygen atoms, giving a 'delocalised cloud' of electrons that confers symmetry and stability. It can be drawn like this, giving pyruvate something of a Cheshire Cat grin:

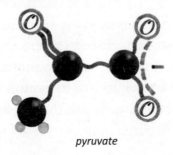

pyruvate

In cells, carboxylic acids are normally found in this negatively charged form, which changes its name to end in 'ate'. Pyruvic acid (with a proton) becomes 'pyruvate' after it has lost the proton. The same convention applies to all carboxylic acids. Notice too that the carboxylate group has the formula CO_2, and it can indeed break off relatively easily as CO_2 (in this case leaving a C2 carboxylic acid, acetate, or vinegar). But for the most part the Cheshire Cat grin says it all: carboxylate groups are as lazy as fat cats, and prefer not to react at all. Overall, pyruvate is quite a complex composite character – a stout, mad-eyed pirate with a smug grin. Pyruvate the pirate.

The fate of pyruvate

Having now familiarised ourselves with the nature of carboxylic acids, let's return to Krebs, who was wrestling with their fate. We've seen that pyruvate itself plays a central role in respiration, yet Krebs knew that other carboxylic acids were involved too. The strange thing was that virtually all of them had *more* carbons than pyruvate, between four and six carbon atoms in length, compared to just three carbons in pyruvate. Why would that matter, I hear you say? Recall that respiration converts organic molecules into CO_2 and H_2O. Glucose starts with six carbons and ends up as six molecules of CO_2. If each carbon is extracted individually, then the remaining molecule must get progressively shorter. Krebs knew that the six-carbon glucose is initially split into two three-carbon molecules of pyruvate, so a reasonable working assumption was that the intermediates should be stripped down to two carbons and then one carbon. In other words, the C3 pyruvate should be broken down into the C2 acetate and ultimately the C1 CO_2. Yet it seemed the opposite happened. Instead of getting shorter, the C3 pyruvate somehow became a C4 or even a C6 carboxylic acid. Why would the carbon chain get longer instead of shorter? It made no sense.

The first clue came from a madcap Hungarian scientist who had completed his PhD with Sir Frederick Gowland Hopkins in 1927. From a family of minor nobility with a scientific bent, Albert Szent-Györgyi was a true original. His mother was an opera singer who had auditioned with Gustav Mahler, then a conductor at the Budapest Opera; Mahler had advised her to marry instead, as her voice was not quite good enough. The young Szent-Györgyi fought in the Great War and became so disgusted after two years that he shot himself through the bone in his arm, claiming that he was wounded by enemy fire – a story he narrates in his short autobiography *Lost in the Twentieth Century*. He completed his medical degree during his convalescence. After some scientific peregrinations around Europe, he eventually arrived in Cambridge, where he began work on what became known as vitamin C; his isolation of this won him a Nobel prize in 1937. Vitamin C is a C6 molecule, quite similar to glucose, with the formula $C_6H_8O_6$. Szent-Györgyi struggled to determine its structure. With an impish sense of humour, he first named it 'ignose' (the 'nose' denoting its sugary affinities, and the 'ig' his own ignorance). When that name was rejected, he proposed 'Godnose'; it was eventually given the chemical name ascorbic acid, for its anti-scorbutic properties (preventing the deficiency disease scurvy).[5]

Krebs and Szent-Györgyi had crossed paths on several occasions when Krebs was still in Germany, and despite some scientific disagreements plainly held each other in high esteem. Indeed, Szent-Györgyi wrote to his old mentor Sir Frederick, bringing Krebs's plight to his attention. Now back in Hungary, Szent-Györgyi published a striking finding in an unusual system – minced pigeon-breast muscle – which he had learned from Dorothy Needham while in Cambridge. This is a flight muscle,

---

5 Szent-Györgyi had a way with words. Some of the most arresting aphorisms in biochemistry trace back to him. One of my favourites is 'Life is nothing but an electron looking for a place to rest.' Szent-Györgyi wrote a number of books, ranging widely across science and life, including a collection of deeply troubled essays on the nuclear threat in the 1960s entitled *The Crazy Ape*, with overtones of Dr Strangelove. I stumbled across one of his books that I didn't know while I was writing this book: *The Living State*, published in 1972. His thinking was astonishingly close to my own, albeit we've learned a lot in the last half century. He begins: 'Every biologist has at some time asked "What is life?" and none has ever given a satisfactory answer.' That's still true today. He continues: 'Life, as such, does not exist. What we can see and measure are material systems which have the wonderful quality of "being alive". What we can ask more hopefully is "What are the properties which bring matter to life?"' I could have written that in my own introduction. Later, there's a celebrated passage, which again resonates with me: 'My own scientific career was a descent from higher to lower dimension, led by the desire to understand life. I went from animals to cells, from cells to bacteria, from bacteria to molecules, from molecules to electrons. The story had its irony, for molecules and electrons have no life at all. On my way, life ran out between my fingers. The present book is the result of my effort to find my way back again, climbing up the same ladder I had so laboriously descended. Having started in medicine, it is befitting that I should end with a medical problem, cancer, which took away most of what was dear to me.'

which can support exceptionally high metabolic rates, because flying is energetically highly costly. That was important at the time because higher rates of respiration made it easier to measure any changes; in retrospect it was also misleading, for reasons that we'll come to. Szent-Györgyi found that adding small amounts of C4 carboxylic acids, notably succinate, dramatically increased the rate of respiration (by up to 600 per cent!!) over prolonged periods. Yet when he analysed exactly how much of the carboxylic acid remained in the muscle after the experiment, he found that *none* had disappeared. These simple molecules sped up reactions without being consumed themselves. That is, by definition, a *catalyst*. It might appear strange, Szent-Györgyi remarked, to attribute to a substance as simple as succinate the role of a catalyst. But what else could it be?

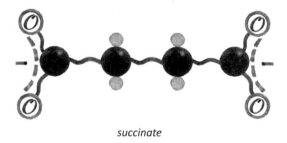

succinate

A catalyst! That comment might not trouble a chemist much, but it was certainly unexpected for most biochemists. Say 'catalyst' to a biologist in a word-association game, and they will probably reply 'enzyme'. An enzyme is a protein, encoded by a gene, typically composed of a chain of several hundred amino acids with a total of perhaps 1,500 carbon atoms. An enzyme has a complex shape and can speed up a specific biochemical reaction with sensational selectivity and speed. In general, one enzyme catalyses a single reaction, being honed to the shape and charge of a specific molecule known as its substrate. But succinate is not like that. It's a simple molecule with just four carbon atoms and nought else beyond the usual oxygen and hydrogen atoms. The molecule is beautifully symmetrical, as my portrait shows. It has *two* carboxylate groups, one at each end – two smugly unreactive Cheshire Cat grins. No mad-eye oxygen atoms. A quick personality read-out says that it is more like an extra-smug banker than a pirate. Yet somehow succinate doesn't increase the rate of a single reaction, but rather the overall speed of respiration. That only made the individual steps even more perplexing. The bottom line is that succinate was fairly unreactive yet had a general catalytic activity, unlike almost anything else known in biology at the time. Almost …

Pulling hydrogen

Szent-Györgyi had a hypothesis. It's an excellent example of how an ingenious hypothesis can drive science forward down a blind alley. Luckily not for long in this case. The idea was this. Succinate was one of four C4 carboxylic acids (the others being fumarate, malate and oxaloacetate), all of which could speed up respiration to a comparable degree, as Szent-Györgyi had shown. All were known to be interchangeable, and none of them was consumed by respiration. They differed from each other in one crucial respect: the number of hydrogen atoms in the short carbon chain. Each differed from its neighbour by two hydrogen atoms; let's adopt the simpler convention of saying 2H.[6] Here they are:

---

6 Let me just clarify the distinction between 2H and H_2. It's an important difference but might leave you feeling exasperated with pedantic scientists. 2H refers to two hydrogen atoms, which are typically attached to other molecules, such as the carboxylic acids shown in the figure. A hydrogen atom comprises a single proton in the nucleus along with a single electron around the nucleus.

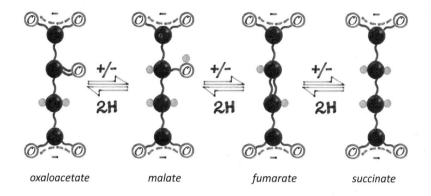

oxaloacetate   malate   fumarate   succinate

As Szent-Györgyi observed, the four C4 carboxylic acids differ from each other by 2H. Notice that there is also an O difference between fumarate and malate, so the difference here is actually H₂O.

Szent-Györgyi had seen this pattern before. Vitamin C can pick up 2H, and then pass them onto something else, notably vitamin E. And Szent-Györgyi had been fascinated by the pale-yellow pigment found in cells that he called 'flave' (which is now known as flavin adenine dinucleotide, or FAD, and is also involved in respiration). It too could lose 2H and in doing so became colourless; on picking them up again it would regain its yellow colour. These spooky colour changes as hydrogen was transferred had entranced Szent-Györgyi. Now here were even more intriguing examples of hydrogen transfer, somehow linked to respiration.

2H therefore refers to two protons and two electrons, which can be drawn from almost anywhere. In contrast H_2 is a gas: two hydrogen atoms bound together by a covalent bond, in which each hydrogen atom shares its single electron, so each atom now has a full inner shell of two electrons, giving quantum mechanical stability. H_2 is also composed of two protons and two electrons, but in this case only in the gaseous form H_2.

Recall that respiration consumes oxygen by reacting it with hydrogen, or more specifically 2H. The product is water, or H_2O. Warburg had already shown that the enzyme that catalyses this final step is the haem protein cytochrome oxidase, but it had to get its 2H from somewhere. Szent-Györgyi pictured a shuttle of carriers that stripped 2H from sugars, and then passed them from one to another, and eventually to oxygen to form H_2O. It was a beautiful conception, and as it turned out, *specifically* wrong. Specifically, but not generally – he was correct that 2H carriers are vital in respiration, but wrong about the identity of the carriers. He'd been blinded by his infatuation with hydrogen carriers. But Krebs wasn't. He saw something else.

By now, in 1935, Krebs had moved to Sheffield. Hopkins had been unable to raise the funds to keep paying his salary in Cambridge, and although he saw Krebs as his anointed successor, there was little he could do in the short term. Ironically, soon after Krebs had gone, a lectureship became available in Cambridge, which Hopkins offered to Krebs, along with the title 'Director of Research'. Krebs was torn, not surprisingly, but he appreciated the support of his new colleagues in Sheffield, where he also had more space. He'd also come to love the wild countryside of the nearby Peak District. He eventually stayed for nineteen years, marrying a local girl and raising a family, before moving to Oxford in 1954.[7]

---

7 He married Margaret Fieldhouse in 1938. She taught in a Catholic convent school near the university, having been instructed in the Montessori teaching method by the inspirational Maria Montessori herself. When Hans asked Margaret's father for her hand in marriage, Mr Fieldhouse said to Margaret 'I don't know what you see in him. The chap seems to have no vices.' Margaret was reportedly amused and answered 'I expect I shall find them before long.' They had two sons and a daughter, the youngest, John, becoming a distinguished zoologist and science policy advisor (chairing the Food Standards Agency, British Science Association and Committee on Climate Change among others).

While stimulated by Szent-Györgyi's findings, Krebs could also see some problems. It boiled down to whether the C4 carboxylic acids were simply hydrogen carriers, as argued by Szent-Györgyi, or were actual *intermediates* in the breakdown of sugars, as Krebs himself thought. This is such a fine example of how clever scientific reasoning works it's worth chewing over briefly. Two abstruse and superficially similar ideas in fact generate opposing predictions that can be tested in the lab. The experiment pointed Krebs straight to the beautiful conception of a cycle, which has played a central, if deceptive, role in biochemistry ever since. It's also worth noting that, even though Szent-Györgyi's idea turned out to be wrong in detail, it was valuable nonetheless, because it focused attention on a primary feature of the cycle – the removal of hydrogen (2H) from carboxylic acids. Again, I have to emphasise just how subtle our reasoning needs to be to understand anything in biochemistry; at no point do we just 'look' at something. Let's think this through.

If C4 carboxylic acids were merely hydrogen carriers, then as each 2H was stripped out from the sugar and passed onto a carboxylic acid, then CO_2 should be released from the skeletal remains of the sugar. Remember that the formula for glucose is $C_6H_{12}O_6$. Szent-Györgyi imagined that all the hydrogens were stripped out from glucose to give six sets of 2H, which were transferred onto the C4 carboxylic acids before being passed onwards to oxygen to form water. Carboxylic acids were acting as shuttles. In the meantime, the carbon husks of the glucose molecule had to be discharged as CO_2.

That makes a clear prediction: adding an excess of the

He was ennobled in 2007 as Lord Krebs and – as mentioned earlier – founded the Sir Hans Krebs Trust in 2015, to support refugee scientists.

carboxylic acid shuttle should lead to the complete break-down of the sugar even in the absence of oxygen. To grasp this, remember that oxygen is normally the final 2H acceptor; it accepts 2H to form water, which is excreted as waste. Now, if the C4 carboxylic acids simply acted as a shuttle of 2H to oxygen, then adding a lot more of them to a tissue slice should mop up all the 2H stripped out from sugars without needing to transfer them onto oxygen. Think of the chain of 2H carriers as a bucket brigade: adding more buckets means that more 2H can be stripped from food, accumulating in the buckets, even if the 2H buckets aren't emptied onto oxygen. We just have more full buckets. If so, then CO_2 formation should be the same, regard-less of whether oxygen was present or not. But the experiment showed that in fact very little CO_2 was formed in the absence of oxygen.

While both men agreed that this was the case, Szent-Györgyi ascribed the discrepancy to issues with measurement in a tricky experimental system, while Krebs interpreted the result literally – he saw the C4 carboxylic acids as *intermediates in* the breakdown of sugars, hence they had not yet had all their CO_2 stripped out, and the final steps to do so required oxygen. Going back to the case in point, if the C3 pyruvate really was an intermediate, as Krebs imagined, then it had to be trans-formed somehow into the next known intermediate – the C4 succinate. That is ostensibly the opposite of breakdown, as the chain length gets longer rather than shorter. Perplexing.

But Krebs had a clue, which others may have overlooked as irrelevant. The C6 carboxylic acid citric acid, which is responsi-ble for the tang of citrus fruit, had recently been shown to break down in several steps to give the C4 succinate. That gave quite a long series of apparently interconnected steps, which ran from the C6 citrate via a C5 intermediate (called α-ketoglutarate) to the four C_4 carboxylic acids. It had not escaped Krebs's notice that these steps released two CO_2 molecules, as the carboxylic acids were clipped down from six to four carbon atoms in length. Presumably, these steps, in which CO_2 was pulled out, needed oxygen to go ahead. Krebs also found that adding citrate to his pigeon-breast slices (he'd modified Szent-Györgyi's method) sped up respiration in a similar way to succinate. Citrate was definitely involved in respiration. In a *coup de grâce*, Krebs showed that if he added citrate to the tissue slices along with an inhibitor of respiration, then the succinate now accumulated. That could only mean that the pathway from citrate to succinate and onwards was indeed a normal part of respiration – blocking respiration led to the accumulation of these intermediates. In short: Krebs now had a succession of intermediates, which somehow went from pyruvate (C3) to citrate (C6) then α-ketoglutarate (C5) and finally succinate (C4). Adding any of these intermediates had a catalytic effect, speeding up the overall rate of respiration. It all added up to pretty much the scientific equivalent of a cryptic crossword puzzle.

Circular reasoning

Only Krebs was thinking about the cryptic crossword in the right way; even his closest collaborators had little inkling of what was on his mind. But Krebs had thought about the catalytic effects of small molecules before. His work on the urea cycle was just that. Unlike anyone else in the field, who all had just the same information, Krebs was already thinking about a cycle.

The urea cycle starts with a C6 amino acid called arginine. An enzyme splits this into two unequal pieces, the C1 waste product urea, plus a shorter (C5) amino acid called ornithine.

What happened to ornithine next was unknown, but Krebs had found that adding more ornithine to his tissue slices somehow increased the rate of urea formation without the ornithine being consumed itself. That had initially baffled him, as normally the accumulation of products in a reaction tends to get in the way of forming any more, yet this did the opposite. Finally, Krebs realised that ornithine was acting as a catalyst: it sped up the reaction, without being consumed itself. To do so, ornithine had to give rise to more arginine, from which more urea could form in turn, thereby regenerating the ornithine he had added. In other words, it had to be a cycle! Arginine was re-formed via several steps from ornithine, through the addition of a CO_2 and two molecules of NH_3. It was this beautiful insight, with just the degree of unexpectedness to be expected of living systems, that Hopkins had celebrated at the Royal Society in 1932.

Krebs realised that biochemical cycles are necessarily catalytic. In essence, adding more of any constituent gives rise to more of the next, and so on in turn, until the cycle has run full circle and regenerated the first constituent again. That explains why none of the constituents are consumed: each one has to be regenerated from its precursor indefinitely. This might sound suspiciously like a perpetual motion machine that operates with 100 per cent efficiency. But of course in reality the cycle must be continuously fed with materials to keep on spinning. If, in the first few steps of a cycle beginning with citrate, two CO_2 molecules and multiple sets of 2H are stripped out, then one full turn of the cycle would need to replenish those same constituents, or it would not be a cycle at all. But the carbon, hydrogen and oxygen that have to be replenished do not need to come in the same molecular form: they could be in the form of an organic molecule combining them all. To return to our traffic analogy from the Introduction, you might think of the Krebs cycle as a fully laden car-carrier arriving at a magic roundabout. At each junction, there's a blinding flash and one of the cars vanishes from the trailer, setting out on its own shape-changing journey, or being scrapped for the metals. Only the tractor unit ever makes it all the way round. Then: bang! The truck is magically reloaded with a full trailer. The cycle can start again only if the tractor unit is coupled to a fully loaded trailer, so each turn of the roundabout always regenerates the tractor. To put that back into chemical terms, Krebs realised that the cycle could be fed with organic molecules (the trailer), and a full turn of the cycle would break them down into CO_2 plus multiple sets of 2H to be fed into the furnace of respiration, leaving only the simplest carbon skeleton (the tractor unit) to be reloaded with the next organic molecule for breaking into its constituent parts.

If so, then which organic molecule was being fed into the cycle? Krebs focused on pyruvate. Remember that pyruvate is the intermediate breakdown product of sugars and some amino acids. Was it being fed into a cycle and catalytically broken down into CO_2 and 2H?

The crux of the matter was the final C4 carboxylic acid, known as oxaloacetate. Could this be the 'tractor unit'? Oxaloacetate was the last clear step in the breakdown of citrate, but its metabolic fate was ambiguous, or rather, it frayed into several possible pathways (we'll come back to these in Chapter 5). Just one of those ambiguous paths mattered to Krebs. It had recently been shown that citrate could be made from oxaloacetate and pyruvate, in effect welding them together under quite harsh chemical conditions, and without an enzyme. If a similar reaction did operate under normal metabolic conditions, catalysed by an enzyme, then Krebs had completed a first draft of his cycle: pyruvate was fed into the cycle by cobbling it onto oxaloacetate, equivalent to loading a full trailer onto the tractor unit. It was a quick experiment to do, and it worked as well as he must have most fervently desired. Adding pyruvate and oxaloacetate to tissue slices did indeed generate citrate quickly under normal physiological conditions. The rest was accounting. Could the proposed cycle account for the consumption of oxygen, the release of CO_2, the loss of pyruvate, the formation of citrate? Yes! Yes! Yes! Yes! It did account for each in full! Here is Krebs's first simple conception of the cycle, more or less as he depicted it in his famous 1937 paper:

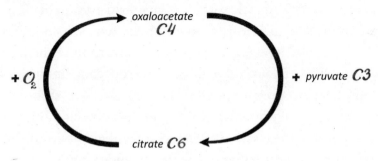

The overall effect of this cycle is to burn up one molecule of pyruvate, or at least something closely related to it. Exactly what was joined onto oxaloacetate to form citrate was still a mystery. It could not have been the C3 pyruvate itself, as depicted here, because that has one carbon too many – one C4 plus one C3 should give a C7 molecule, not the C6 citrate. One carbon was presumably lost as a CO_2 somewhere down the line. Perhaps this uncertainty was partly behind one of *Nature*'s more famous rejections – the journal returned Krebs's paper on the cycle with a quaint recommendation that he might prefer to publish it elsewhere as they had a long queue of papers awaiting publication.

A full decade later, in 1947, Fritz Lipmann discovered that pyruvate was indeed stripped of one CO_2 (and 2H) leaving a

C_2 carboxylic acid attached to a larger molecule that acted as a kind of molecular handle. He named the overall assemblage *acetyl coenzyme A* or acetyl CoA for short. This turned out to be perhaps the single most central molecule in metabolism across the whole of life, from ancient bacteria to ourselves, and we will encounter it repeatedly. Once formed, acetyl CoA reacted with oxaloacetate to make citrate. So the 'full trailer' was really acetyl CoA.

Lipmann had trained in Berlin in the late 1920s, in the lab next door to Krebs. He, too, later fled from the Nazis, in his case from Denmark to New York and eventually Boston. Lipmann was a notoriously slow thinker, who would interrupt and ask people to repeat what they had just said, yet he seemed able to read between their lines, interrogating the unspoken assumptions to see things that nobody else had spotted. He pursued a lifelong interest in the energy currencies of biology, and the ways they are exchanged. Lipmann's extraordinary intellectual bequest to biochemistry included what he dubbed the 'universal energy currency' of life, ATP (short for adenosine triphosphate) along with acetyl phosphate and acetyl CoA itself, three central pillars of energy metabolism. We will return to all three of them when we think about the origin of life in Chapter 3, following in Lipmann's own footsteps.

The term 'acetyl' refers to the 2C acetic acid – vinegar – which is not very reactive. Look at my portrait of it below and you can see that it has the stout belly and Cheshire Cat grin of pyruvate – but no mad-eye oxygen to stir up trouble. To make it react with oxaloacetate, the acetate needs to be activated by attaching it to coenzyme A. Lipmann didn't designate coenzyme 'A' to distinguish it from some coenzyme 'B'; the 'A' stood for 'activation of acetate'. We can put aside the baroque complexity of coenzyme A here, and simply consider the acetyl group in my portrait. Note that one of the oxygen atoms in acetate is replaced with the sulfur of CoA (which is marked with an 'S'); this slippery sulfur is what makes the composite molecule more reactive than acetate itself (and that remaining oxygen is beginning to stare a bit):

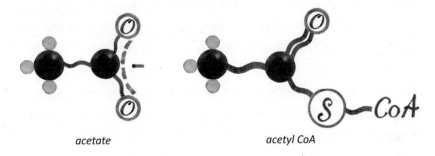

acetate   acetyl CoA

Once activated in this way, the C2 acetate can be attached onto the C4 oxaloacetate, to give the C6 citrate. At last, everything added up properly. The process by which glucose is oxidised goes like this. First, the C6 glucose is split into two molecules of the C3 pyruvate, each of which is broken down to acetyl CoA. These are then fed into the Krebs cycle. One complete spin of the cycle (from pyruvate) generates three molecules of CO_2 and five sets of 2H, equivalent to five molecules of H_2.[8] This hydrogen is then fed to oxygen to generate energy, in the form of ATP, via cellular respiration.

There is some poetry in the parallel lives and eventual convergence of the scientific ideas of Lipmann and Krebs, so it

---

8 For the more astute accountants among you, some of the hydrogens burnt in fact derive from water, which is added at two separate steps onto Krebs-cycle intermediates (see figure on page 286 for one of them). Recall that stripping the 2H from glucose would give you six sets of 2H, whereas the Krebs cycle actually strips out ten sets of 2H. So nearly half the 2H burnt in respiration comes from splitting water (which is basically what happens in photosynthesis) and then burning the extracted hydrogen. Did you know you could do that? Just wow!

seems only fitting that they shared the Nobel prize in 1953. Yet this was by no means the end of the story. There was another detail (one might call it the whole point) which Krebs could not explain in 1937; nor could Lipmann in 1947. The question taunted: how was the energy released by burning multiple sets of 2H captured by cells and used to drive metabolism? Although Lipmann had shown that the energy released was conserved in the form of ATP, nobody knew how the burning of 2H was coupled to the synthesis of ATP. It was the burning question in biochemistry over the next two decades. And it took a very different genius nurtured in Hopkins's Cambridge laboratory to solve the problem.

The living structure of cells

Peter Mitchell arrived as an undergraduate in Cambridge several years after Krebs had left, at the outbreak of war in 1939. A sporting injury at school had prevented him joining the army. He had been a mixed student, and his example serves as another reminder of the dangers of categorising intelligence. He had flourished in maths and physics, subjects that he could understand from the bottom up by working out the underlying principles for himself; but he was dismal at English and history, where he couldn't see any fundamental principles at all. He only scraped into Cambridge because his school's headmaster, a mathematician himself who could see how gifted Mitchell was, had intervened on his behalf.

Mitchell hardly repaid his trust. He did badly in his degree and initially failed his doctorate. He had written an opaquely philosophical thesis on how bacteria transport molecules in and out of the cell, grounded in his own development of ancient Greek theory on 'fluctids' and 'statids'. The examiners dismissed his work as 'not really a thesis at all'. Mitchell's mentor, the distinguished Polish biochemist David Keilin, commented that 'Peter is too original for his examiners.' He finally passed his PhD in 1951, with a more conventional thesis on the mechanism of action of penicillin. Like Francis Crick, who took seven years to complete his own PhD thesis, having taught himself the mathematical theory of X-ray crystallography along the way, Mitchell might not have thrived in today's research environment.

Despite these failings, Mitchell certainly impressed those around him. Fred Sanger, a double Nobel laureate, wrote that 'Peter had an original idea on every subject and we all knew even then that he would possibly change science.' Just after the war, in 1947, Marjory Stephenson was elected President of the Society for General Microbiology (with strong support from Alexander Fleming and J. B. S. Haldane). In this capacity, she invited Mitchell to deliver a keynote lecture at the Society's annual conference – an unprecedented honour for a graduate student. Tragically, Stephenson died of breast cancer before the meeting took place. In one of her final acts, she recommended that her laboratory assistant Jennifer Moyle should work with Mitchell, thinking the pair would complement each other well. Moyle was a brilliant experimentalist, while Mitchell was brimming with radical scientific ideas, but was less focused in the lab; a fellow student commented that Peter wanted to debate what was going on, rather than get on with the prescribed experiment.

I can hardly begin to imagine what Jennifer Moyle might have accomplished in another age. She had gone to Girton College Cambridge in 1939 where she studied Natural Sciences, specialising in biochemistry. She was inspired by the lectures of the comparative biochemist Ernest Baldwin, another protégé of Hopkins, while attending many lectures in philosophy. She completed her 'Title of Bachelor of Arts' degree in 1942 – sad to say, formal degrees were not awarded to women in Cambridge until 1948, when Girton College was granted full college status by the university. Straight after her 'title of' degree, Moyle joined the Auxiliary Territorial Service, the women's branch of the British Army during the war, and she soon went into military intelligence, becoming an intelligence officer in MI8, the group responsible for signals intelligence. By the end of the war she was second in command of a unit that analysed intelligence from the breaking of German ciphers, and afterwards spent a year helping prepare servicemen for their return to civilian life. She then joined Marjory Stephenson in Cambridge as a research assistant in 1946, first meeting Mitchell at departmental tea in 1947. She was struck by his keen intellect and wide-ranging interests, as well as his cultivated 'Beethovenian' appearance, with a flowing mane of hair (he too later went deaf). Moyle herself was an enthusiastic musician, who sang in choirs throughout her life.

Marjory Stephenson could hardly have been more prescient in recommending that Mitchell and Moyle work together. They formed a lifelong scientific partnership, which drove a paradigm shift in twentieth-century biology, starting in the free air of Hopkins's biochemistry department, before moving to Edinburgh in the mid-1950s, and culminating at the Glynn Institute, the remarkable eighteenth-century manor house near Bodmin in Cornwall that Mitchell lovingly restored as a laboratory and home.[9] The Glynn was animated by Mitchell's unique vision of

---

9 Mitchell had some wealth through his uncle, Sir Godfrey Mitchell, a construction engineer who had bought a small, insolvent construction company, George Wimpey, in 1919. Under Godfrey's dynamic direction, Wimpey went on to build some 300,000 homes in England by the time of his death in 1982. During the Second World War, the Wimpey Construction Company built hundreds

science, and became a place of pilgrimage for bioenergeticists the world over, who would stay for weeks or months to work in the lab, discuss the physics of biology, and drink deeply from the source.

For source it was: Mitchell reconceived energy flow, in a way that we are only just beginning to assimilate properly into biology today – a theme I'll return to later in the book. Krebs had thought beyond linear pathways and discovered cycles, four of them in all. But Mitchell really transcended chemistry altogether, and talked in a new language that few understood, about the proton-motive force, electrical membrane potentials, vectorial chemistry and proticity. While Mitchell's theories were dressed in mysterious mathematical symbols and rejected the crude notion of the cell as a bag of molecules dissolved in water, his ideas were quite simple in essence. They grew from his early preoccupation with how bacteria keep their inside different from the outside. Mitchell realised that bacteria had to recognise specific molecules with the same finesse as an enzyme, and then actively transfer them in or out of the cell across its bounding membrane – they had to pump. A selective pump embedded in the membrane plainly needs to be powered, which costs energy. Conversely, Mitchell saw that if the pump went into reverse, allowing whatever had been pumped out to flow back into the cell down a concentration gradient, then the energy released could in principle be put to work. The pump becomes a turbine. This is no more complex than blowing air

---

of airfields, balloon stations, docks and army camps, for which Godfrey was knighted in 1948; it built Heathrow Airport soon afterwards. During the war years, the success of Wimpey allowed Peter Mitchell to cut a flamboyant figure around Cambridge, driving a silver Rolls-Royce, with his flowing mane of hair. Later on, periodic gifts of shares allowed him to purchase and renovate Glynn House, and helped him to maintain the institute through difficult times.

into a balloon. Releasing the pent-up air can power work, in this case zipping the balloon around the room by jet propulsion.

Warburg and Krebs had both talked about the 'living structure of cells', arguing that some processes were 'bound to the life of the cells'. Many biochemists were uncomfortable with this view, as it recalled the discredited idea of vitalism, implying that there was something special about living matter that could not be reduced to mere chemistry; and frankly, this view is still prevalent today. The goal of biochemistry was to grind up tissue, isolate an enzyme, then measure its precise function in as pure and uncontaminated a preparation as was humanly possible. Fermentation was the classic example: it did indeed proceed in homogenised tissues. Other biochemical reactions did not work initially, but with better preparation methods were eventually demonstrated in cell-free extracts. So, while Warburg and Krebs contended that respiration could only take place in intact cells, most biochemists thought it was only a matter of time and more meticulous methods before all biochemical processes could be demonstrated in purified preparations.

For Mitchell that view was anathema. The integrity of the membrane was indispensable to the life of cells. The membrane is the thin oily layer that surrounds all cells, barely six millionths of a millimetre (6 nanometres) in thickness, enclosing the gel-like cytoplasm inside. Some cells such as plants, fungi and bacteria also have a more robust, mesh-like cell wall, which prevents swelling and protects against mechanical injury, while allowing the movement of small molecules in and out; but it is the flimsy membrane that matters here. Being composed of oily lipids, the membrane obstructs the transfer of charged particles, even tiny protons. If critical membranes sprang a leak to protons, the cell would swiftly die. Mitchell's profound insight was to see that this law applied not only to the transport of materials in and out of bacterial cells, but to the process of respiration itself. For 'the living structure of cells', read 'membrane'. To burn the 2H derived from the Krebs cycle, a membrane was needed. That observation alone was enough for Mitchell to frame his whole hypothesis. We'll see that the Krebs cycle is intimately linked to the electrical properties of this membrane, just as Mitchell predicted – but in ways that neither Mitchell nor Krebs could have begun to imagine.

Separating charge

I have talked about stripping hydrogen (2H) from molecules and feeding them to oxygen. These 2H are not free in solution, nor are they shuttled by carboxylic acids, as Szent-Györgyi once thought; it turned out they are bound to a large molecule called NAD^+ (or 'nicotinamide adenine dinucleotide', since you ask). The form of NAD^+ that has picked up the components of hydrogen (or more exactly, two electrons and one proton, leaving another proton nearby, bound onto water) is called NADH.[10] NADH will figure large later in this book, but for now, when you see the term NADH, just focus on the 'H'. I hope you will develop a Pavlovian response, and when you see NADH, you'll think 'Aha! Once more this is the beast that transfers hydrogen, or rather, the two electrons and two protons that constitute 2H!'

---

10 NADH was also discovered by Warburg, but he failed to appreciate its full significance. It fell to the great US biochemist Albert Lehninger (author of a famous textbook) to show that the oxidation of NADH by oxygen was linked to ATP synthesis, in two classic papers in the late 1940s. Lehninger's pioneering work on bioenergetics included demonstrating that respiration takes place in mitochondria, the 'power-houses' of the cell. Yet for all his brilliance, Lehninger could not solve exactly *how* NADH oxidation was coupled to ATP synthesis. The entire field got stuck for two decades. It took Mitchell's unprecedented conceptual leap to answer the question.

In cell respiration, NADH transfers its 2H to oxygen to form water (H_2O), capturing some of the energy released in the process. But NADH does *not* transfer its 2H directly to oxygen; instead, the electrons hop (in actuality, they 'quantum tunnel') down a chain of carriers associated with the membrane itself, what's now called the respiratory chain – an idea conceived by Mitchell's mentor David Keilin.[11] Notice that I say *electrons*, not whole hydrogen atoms. What happened to the protons was anybody's guess, though plainly they too ended up on oxygen to form water. Only Mitchell saw the path of protons as the key to the decades-old puzzle of coupling – how exactly the transfer of 2H to oxygen is coupled to the synthesis of ATP.

Mitchell proposed that the 2H (derived from the Krebs cycle, in the form of NADH) are split into their component protons and electrons. The electrons are transferred to oxygen by way of the respiratory chain of carriers embedded in the membrane itself – an electrical current, insulated by the surrounding lipids. This electrical current powers the extrusion of protons across the membrane. Mitchell knew that the membrane is nearly impermeable to protons; any damage to the membrane that renders it leaky to protons will dissipate respiration. The protons accumulate outside, giving a difference in proton concentration (which is to say, pH) between the inside and outside. Critically, because protons are positively charged, their accumulation outside generates an electrical

---

11 A much-loved figure, with a faintly owlish look, Keilin headed the Molteno Institute in Cambridge, just across the lawn from Hopkins's biochemistry department. He had welcomed Krebs into his home a decade earlier, despite a long and acrimonious dispute with Warburg about the nature of respiration. The basis of that dispute was three separate cytochromes, each one distinct from Warburg's ferment, which Keilin had detected and proposed acted as a 'respiratory chain' of carriers that passed electrons to oxygen.

charge across the membrane, analogous to a battery. Finally, the flow of protons back through protein turbines embedded in the membrane powers the synthesis of ATP, Lipmann's universal energy currency. So ATP synthesis is powered by what Mitchell called the *proton-motive force*. Only then are protons reunited with electrons on oxygen, to form water.[12] Overall, it looks something like the figure overleaf.

All this is now received wisdom, but it was hard won. These ideas were bitterly contested over two decades. Jennifer Moyle's experiments contributed largely to the eventual acceptance of Mitchell's ideas, for which he alone was awarded the Nobel prize in 1978. Her essential contribution would surely not have been overlooked today. Mitchell and Moyle together published a series of ground-breaking papers in *Nature* in the mid-1960s, laying out the basic experimental approaches still used today. Let me just give you a flavour (don't worry about the details). They demonstrated that mitochondrial membranes really do pump protons when given various substrates (including succinate) and oxygen, transiently acidifying the medium. They showed that inhibitors of electron transfer, such as the pesticide rotenone, partially blocked the acidification of the medium. And they showed that 'uncouplers' such as the antibiotic

---

12 There's an irony here because Krebs and R. E. Davies came up with a similar idea a decade before Mitchell, in 1951, in which they proposed that 'It is feasible that ionic concentration differences form the link between the free energy of respiration on the one hand and ATP synthesis on the other.' They went on to say 'the operation of a mechanism depending on concentration differences necessarily postulates specific structures which prevent wasteful mixing of ions' – a membrane. But Krebs and Davies never went into testable detail, still less testing it, and never discussed electrical membrane potentials. Mitchell had been unaware of Krebs's idea, and some years later apologised to him for overlooking it; Krebs responded: 'After all it was no more than an idea – hardly a hypothesis – and did not lead directly to any useful experiments.' What a magnanimous man.

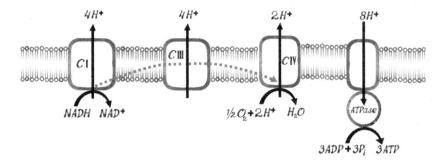

*A grossly simplified schematic of the respiratory chain.
The transfer of electrons from NADH to oxygen is shown
by the dotted line. These electrons pass through three great
protein complexes embedded in the membrane (complex
I, III and IV; we'll meet complex II in Chapter 4). This
current of electrons powers the extrusion of ten protons
across the membrane, generating a combined electrical
charge and concentration difference (the proton-motive
force) that drives ATP synthesis from ADP (adenosine
diphosphate) and inorganic phosphate (P_i). In humans, eight
protons pass through the ATP synthase for one turn of the
motor, which generates three ATPs. The other two protons
are not 'wasted' but are saved up towards the next ATP.
Note this means that the number of ATPs formed from
a single NADH is not a whole number but a fraction (3.3
in humans) and can vary depending on factors that have
kept bioenergeticists busy for decades, right up to today.*

gramicidin, which disconnect electron transfer from ATP synthesis, allowed protons to flow back through the membrane
instead of driving ATP synthesis. By controlling these experimental parameters with exquisite finesse, Mitchell and Moyle
were even able to establish rough stoichiometries, which is to
say, the ratio of the number of protons pumped to the number

of oxygen molecules consumed, or ATP synthesised. These papers are wholly modern in their thinking and idiom, despite being breathtakingly new in conception.

Such experiments established the indispensable role of the membrane in respiration, as well as the basic topology of proton pumping and ATP synthesis, but said little about how protons were actually pumped. Mitchell had revolutionary ideas here too, although he was wrong about some of the specifics. In particular, he long opposed the idea that pumping could be accomplished by rapid conformational changes in proteins spanning the membrane (rather than in the order and properties of the carriers themselves). In short, Mitchell's beautiful conception was right in principle – electron transfer from $2H$ to O_2 is indeed coupled to a proton circuit that drives ATP synthesis, which Mitchell and Moyle demonstrated together – but wrong about some of the detailed mechanisms of pumping.

Perhaps Mitchell's greatest error concerned the iconic ATP synthase. This extraordinary protein is a rotating motor embedded in the membrane. Dubbed 'the elementary particles of life' when first observed by electron microscopy in the early 1960s, there are many thousands of them in every mitochondrion. We now know that the inward flow of protons through these nanoscale turbines drives the synthesis of ATP. Their operation is wholly mechanical, utterly different from anything that Mitchell had ever imagined, and yet the ATP synthase has come to represent the essence of his hypothesis – its rotation, up to 500 revolutions per second, is powered by the proton-motive force, just as he predicted from the beginning.[13]

---

13 Although Mitchell had himself proposed conformational changes in proteins in his early work on bacteria, he stubbornly resisted the idea in later years,

Regardless of his errors, Mitchell's conception was just magnificent. The energy released by the reaction between 2H (extracted from Krebs cycle intermediates) and oxygen is transduced into an electrical charge on the membrane. This charge is awesome. It might not sound like much – just 150–200 millivolts – yet the membrane is extremely thin, only 6 nanometres in thickness as noted earlier. This means that the electrical field strength (what you would experience as a molecule in the membrane) is about 30 million volts per metre, equivalent to a bolt of lightning across every square nanometre of membrane. And the area of electrically charged membrane in the human body defies belief. If we were to iron out all the mitochondrial membranes in the body, so they were flat, they would cover an area equivalent to about four football pitches – all charged with the power of a bolt of lightning.

This is how respiration works in our own mitochondria, the 'powerhouses' of complex cells. When Mitchell was developing his chemiosmotic hypothesis, in the mid-1960s, Lynn Margulis was beginning to marshal the evidence that mitochondria derived from free-living bacteria that had moved inside other

---

in a long-running dispute with Paul Boyer (who eventually received the Nobel prize alongside Sir John Walker for their work on the structure and mechanism of the ATP synthase). Mitchell's reasoning was curious – it related to testing hypotheses, not the veracity of the idea itself. Mitchell was an ardent supporter of Karl Popper's ideas on the formal need to disprove hypotheses in science. By initially specifying alternating carriers of electrons and 2H, Mitchell also specified precise (therefore testable) ratios of protons pumped to electrons transferred. Simply imagining a protein that changed its shape lacked proper rigour, Mitchell argued, because a conformational change could be associated with the transfer of any number of protons or electrons, making it difficult to disprove a hypothesis, regardless of whether it was true or not. Despite these errors, Mitchell was in fact correct about one of the most baffling pumping mechanisms, known as the Q cycle, in which quinones transfer protons across the membrane through precisely the type of spatial coupling proposed by Mitchell.

cells (in what's called an endosymbiosis) nearly two billion years ago. I've discussed the singular importance of mitochondria to complex life in other books and we can put it aside here. The key point for us right now is that bacteria work the same way as mitochondria, except that their electrical charge is on the bounding membrane of the bacterial cell, as Mitchell knew well – a cellular force field. In bacteria, this electrical force field powers far more than ATP synthesis, explaining in part why this mechanism is universally conserved across all life. Crucially, we'll see that the proton-motive force drives CO_2 fixation in some of the most ancient bacteria, making it relevant to the very beginnings of life. I'll argue that this force field might even cohere cells as individual entities, defining the 'self', if such a thing can be said to exist for bacteria (I think so). I'd like to think that Mitchell would have been sympathetic to this view of life, as it is grounded in his own philosophical conception of biology.

But that's not the message that others have taken from the extraordinary sophistication of cell respiration. Biochemistry turned away from the philosophical ideas that underpinned Mitchell's framework, and concerned itself with the marvellous mechanics of proton pumps and nanomotors. We are left with cellular mechanics that seem to be so Platonically perfect that it's hard to imagine how they might have evolved. At face value, the spinning of the Krebs cycle generates an electrical membrane potential, humming in trillions of our mitochondria, which in turn spins the ATP synthase nanomotors. Cycles driving cycles. Cogs within cogs. No wonder biochemists prefer to avoid talk of origins, or speculations on the evolution of perfection. But that would be to overlook hard prosaic reality. Perfect it ain't.

Bizarre contortions

Imagine you're starving. You are beginning to break down the proteins in your own muscles for the energy you need to keep going. Nothing much can be harvested from the proteins themselves; it's their component amino acids that you need to burn. That's how it began for Krebs back in the early 1930s. Recall that amino acids are split into ammonia plus the carboxylic acids that enter the Krebs cycle. By far the most common amino acid that is shipped around the body from muscle wasting is glutamine. This is converted into the C5 carboxylic acid α-ketoglutarate. The α-ketoglutarate spins round the Krebs cycle to form ... more α-ketoglutarate. Don't think of the Krebs cycle as a furnace burning organic molecules. It's catalytic – and it catalyses the breakdown of the 2C acetate, not the carboxylic acids themselves. So the Krebs cycle can only spin faster if it's provided with more acetate. Normally that comes from the breakdown of glucose or fats; but if we're starving, we have already burned most of them. We need to turn to proteins instead.

Of course we can burn amino acids, but it can be surprisingly complex. Here's what happens in the case of α-ketoglutarate. It is first broken down by way of the Krebs cycle to one of the C4 carboxylic acids, malate. Malate is then exported out of the mitochondria, across two membranes, into the cytosol. There it is oxidised to oxaloacetate, then to an activated form of pyruvate called phosphoenol-pyruvate, and finally to pyruvate itself. Pyruvate is shipped back into the mitochondria, where it is stripped of more hydrogen and CO_2 to form acetyl CoA, and fed back into the Krebs cycle. At last it can be fully burnt. What a convoluted path! It's hard to imagine a rational engineer would have come up with that. Why so roundabout?

Part of the answer relates to contingency. The Krebs cycle and cellular respiration take place inside the mitochondria, which we've noted were once free-living bacteria; they brought along their bacterial metabolism to the ancestor of complex (eukaryotic) cells two billion years ago. To some extent they remain another world within the cell, enclosed by two membranes, the innermost one so impermeable, even to protons, that the only way across is through selective pumps. That's why malate has to be exported and pyruvate imported. But that only answers part of the question. The reactive molecule phosphoenol-pyruvate is a central metabolic branch point, for example in the synthesis of new sugars, which are needed for many reasons, notably making the sugar-phosphate backbone of DNA and RNA. Most of this synthesis takes place outside the mitochondria. So new sugars can be made from amino acids, via the Krebs cycle, and in our own cells this thoroughfare often involves transport in or out of the mitochondria. And that brings us back to a point I touched on earlier – the lucky choice of model made by Szent-Györgyi and Krebs: pigeon-breast muscle.

Adding succinate to pigeon-breast muscle raised the respiratory rate by 600% because that tissue is poised to oxidise glucose as quickly as possible, to provide energy for flying (and the tissue slices were supplied with plenty of glucose needed to form acetyl CoA). But had they repeated the experiment with liver tissue, they would not have seen the same thing. That's because the liver does all kinds of other jobs which require flux in and out of the Krebs cycle, such as making glucose from amino acids. It's not really a cycle so much as a roundabout, where traffic enters and exits as well as going around. Overall there has to be balance, but there is no sense in which it is normally catalytic, however much that particular insight opened the door to Krebs and generations after him. Catalysis is a special case.

The reason is simple enough but has profound consequences for life, which we will explore in this book. The Krebs cycle supplies precursors for the synthesis of amino acids, fats, sugars and more. If Krebs-cycle intermediates are siphoned off for these syntheses, then fewer intermediates remain to provide hydrogen for respiration. Because each intermediate can be regenerated through one complete spin of the cycle, you could think of each molecule as its own mini-cycle, with millions of intermediates in each mitochondrion all whirring around in their own private cycles. Draw off intermediates to make new sugars, and you'll have fewer mini-cycles – a direct sap on the cell's energy, just when you need it most for making new molecules. Add more intermediates, and you'll have extra mini-cycles, each one able to spin off hydrogen that can be burnt in respiration – the catalytic effect observed by Krebs. To keep an overall balance, the intermediates siphoned off for any biosyntheses must be replaced by an input to the Krebs cycle from elsewhere (what's called anaplerosis). So, for much of the time, flux enters and exits the cycle at every junction. All of this passing trade uses the Krebs cycle as a roundabout and not a full cycle. It adds up to no ordinary roundabout, but a crazy traffic system like the Arc de Triomphe roundabout in Paris or the famous magic roundabout in Swindon, with circles within a circle.

It might seem easier to explain the evolution of a crazy roundabout system than a perfect cycle, but these reflections really just raise the opposite problem. Why did several billion years of evolution come up with such a crazy system, or fail to improve it? Krebs himself grappled with this question, and in his final year, 1981, he wrote a paper on the evolution of metabolic pathways. It seems to me that, although he was perfectly aware of the synthetic side of the cycle, he tended to see the cycle in animals as largely dedicated to the breakdown of organics; and that's still how it is presented in most textbooks, and how I have presented it in this chapter: breakdown is its primary purpose, and synthesis is secondary. That impression is strengthened by the fact that key metabolic branch-points such as phosphoenol-pyruvate lie outside the mitochondria, off the Krebs cycle itself, which therefore seems to fall slightly outside the metabolic centre of gravity in animals. Krebs knew that many microbes living in the absence of oxygen use the cycle mainly for synthesis. Yet not once, to my knowledge, did he cite a revolutionary series of papers stretching back to 1966, which showed that in some ancient bacteria the Krebs cycle can spin in reverse – rather than stripping 2H and CO_2 from food to generate energy, it uses energy to react 2H and CO_2 to form organic molecules. In the next chapter, we'll see that this so-called 'reverse' Krebs cycle makes much more sense of evolution.

2

THE PATH OF CARBON

Picture a tree in new leaf, its greens fresh and luminous. Or a giant redwood, towering through the mist, its needles dark and bristling. Dewy grass underfoot, with patches of springy moss. A pond nearby, bursting with reeds or partially concealed beneath duckweed. Perhaps choking with algae. Or picture the savannah, yellowing in the summer haze, with occasional scrubby trees silhouetted against the shimmering heat. Tropical rainforest, dizzyingly high, lianas dangling from the canopy. Red desert, with majestic cacti thrusting against the rocks. Arctic tundra, the last silver birch giving way to reindeer mosses and heathers, if you're lucky flowering exuberantly in spring. Lichens splashing green and orange across the rocks, yet through the microscope a miniature garden from another world. Break open a rock in Antarctica and scrutinise the blue-green film of cyanobacteria beneath the surface. Or think of a luxuriant kelp forest beneath the sea, dazzling rays of light penetrating through darkness that shelters who knows what. Now a coral reef, exploding with colour and pattern, the groping polyps utterly dependent on single-celled algae for their lives and splendour. Try not to think of bleached coral, open graveyards of pointing skeletal fingers, photosynthesisers gone.

These may be clichéd images, but they convey the wondrous

range of photosynthesis on our planet. They summon up landscapes familiar from photographs or films, or if we are fortunate, from exploring ourselves. Either way, plants are powerfully evocative, conjuring up much of what we love about our world, apparently slumbering at peace. They hardly evoke Nature red in tooth and claw, and certainly not the idea of stultifying uniformity. Yet dig beneath the surface, dig into the green of the leaves and you'll find that they are all the same. Close to half the protein found in the leaves of these plants is more or less identical. This is the most abundant protein on Earth, and arguably the most important: it is the protein that takes carbon dioxide from the air and converts it into organic molecules, the living matter of all plants, and ultimately of you and me. It has an enigmatic name – rubisco. Originally an affectionate poke at a pioneer of photosynthesis research, Sam Wildman, on his retirement, the name stuck (it was admittedly better than his own 'fraction-1 protein'). If rubisco sounds more like a breakfast cereal, that was precisely the point; Wildman had spent some years trying to produce a health supplement from his protein (in effect, a cereal) as it is a well-balanced source of essential amino acids.[1] Rubisco is sometimes semi-capitalised as a rather fussy acronym, RuBisCO, which if you insist stands for ribulose bisphosphate carboxylase-oxygenase. Each of those terms has a story to tell, but let's not rush ahead.

Rubisco is so abundant because it is unusually bad at its job,

---

[1] The venture failed because in the late 1970s, when Wildman set up a pilot plant in North Carolina to extract large amounts of rubisco from tobacco leaves, the dangers of smoking were becoming universally recognised. Unsurprisingly, the idea of adding a tobacco protein to food gained little commercial enthusiasm even though the same protein is found in all leaves. Why Wildman persisted with tobacco I have no idea. I presume because his early work was on the tobacco mosaic virus and after that he'd always worked on tobacco plants. It's hard to teach old scientists new tricks.

for an enzyme. Enzymes normally rattle along at incomprehensible rates, catalysing exactly the same reaction hundreds or thousands of times every second. Rubisco, in contrast, has a turnover rate of less than ten reactions per second, meaning that it converts (or 'fixes') fewer than ten molecules of carbon dioxide into organic molecules every second. To increase the rate of photosynthesis, plants are obliged to make more rubisco, hence its excessive abundance. But it's worse than that. Rubisco is also less specific than most enzymes: it can struggle to tell the difference between carbon dioxide and oxygen (which, to be fair, is only one carbon atom different, CO_2 versus O_2). When rubisco first evolved, probably several billion years ago, the CO_2 levels in the atmosphere were far higher than today, while the O_2 concentration was a fraction of its modern 21 per cent. That means rubisco's poor discrimination was no problem back then; it was much more likely to fix CO_2 than O_2. Today the situation is reversed, with O_2 levels hundreds of times higher than CO_2, especially inside actively photosynthesising leaves, where CO_2 is greedily gobbled up and O_2 spews out as the waste product. If the climate is hot and dry, the pores (stomata) that permit gas exchange in leaves start to close, restricting water loss. That gives rubisco a serious problem, because the O_2 waste is trapped inside while CO_2 levels dwindle. Crops can lose a quarter of their yield under these conditions, as they 'fix' more O_2, and then need to work through some elaborate biochemical shenanigans (termed photorespiration) to rectify matters.

Not surprisingly, there has been commercial interest in replacing rubisco with a substitute of ostensibly more rational design, with the carrot of improved yields dangling distractingly overhead. Yet rubisco – which swiftly came to be seen as *the* enzyme for carbon fixation – is emblematic of a more

interesting and deeper problem: the dogma of the unity of biochemistry. In the Introduction, we touched on the value of this
idea; in this chapter, we'll see that it can be malignantly misleading too. Every dogma has its day, but this particular meme
still bedevils the textbooks. The meme is this: photosynthesis converts CO_2 into sugars (using rubisco), while respiration
burns those sugars. To twist a phrase, this is the central dogma
of metabolism: sugars are the backbone of biochemistry.
Their pathways take central position on biochemical charts.
If a conjuror distracts attention from where the real action
is taking place, the teaching of photosynthesis and respiration does exactly that, without even noticing the deception.
The problem lies in our endless fascination with plants and
animals, mirrored in the venerable disciplines of botany and
zoology. What we're taught can condition how we think for
decades. The problem in this case is that focusing too much
on sugars has distracted us from the fundamental connection
between the origin of life, the deep currents of evolution and
our own biochemistry.

The simplistic dichotomy between plants and animals –
the notion that plants make sugars by photosynthesis, while
animals respire those sugars via the Krebs cycle – is of course
partly true, but also profoundly misleading, making little sense
of the structure of metabolism, and therefore little sense of
why it goes wrong in disease. We are taught in school that
plants are 'autotrophs' – they 'make sugars' from inorganic
molecules such as CO_2 and water – while animals eat plants,
or other animals that eat plants, completing the circle of life.
Autotrophs, we're told, depend on the sun for their energy in
the glorious everyday miracle of photosynthesis, transforming light energy and gases from the very air into their own
substance, into our own substance. This is biosynthesis – the

conversion of simple molecules such as CO_2 into the building blocks of life, and ultimately into the great macromolecular machines, DNA and proteins. It is a miracle that should be taught to every schoolchild, but we should tell them too that plants were latecomers on this Earth. For several billion years, our planet was sculpted by bacteria. Photosynthesis as we know it was invented by the cyanobacteria, once graced with the more evocative name blue-green algae. Yet they were quite late to the party too. Before these sophisticated bacteria, other autotrophs were already ancient, living from gases and rocks, sometimes deep in the bowels of the Earth. Many had no need for the sun. If we want to understand the deep chemistry of life (and death ...) we must shift our gaze from the lofty pinnacles of evolution down to the obscure canyons of creation.

In this chapter, we'll see that some of the most ancient bacteria turn the familiar Krebs cycle on its head – they use it in reverse, as a biosynthetic engine that converts the gases CO_2 and H_2 into organic molecules, driving growth. I'll stick with the term 'reverse' here, as these bacteria do indeed reverse the familiar cycle discovered by Krebs, but it makes much more sense to view this biosynthetic version as the forward direction, and our own cycle as the flipped form. Putting terms aside, we'll see that the ancient biosynthetic Krebs cycle was fixing CO_2 a billion years before rubisco and the evolution of photosynthesis in the cyanobacterial ancestors of plant chloroplasts. When it first emerged, the reverse Krebs cycle had little to do with energy generation, instead providing the carbon skeletons needed for biosynthesis. This perspective elucidates the deep metabolism of cells, yet it is still largely missing from the more medically oriented textbooks. It's a serious omission.

To see why that's the case, we first need to follow the path of carbon in photosynthesis, to put rubisco in its proper place. Our winding path will start to straighten out the circular logic of the Krebs cycle – how it came to be the conflicted engine of biosynthesis today. We'll set out with the extraordinary story of the discovery of radioisotopes, notably carbon-14, which underpinned the mythology of photosynthesis. This spanking new approach gave enormously more power than the methods pioneered by Krebs and Warburg, yet it proved as treacherous to interpret. The quest to understand the succession of steps by which CO_2 and water are transformed into the substance of life is fittingly dramatic, and for an everyday miracle, fittingly elusive. Having finally established all the steps, we made the all-too-human mistake of believing them always to be true.

The advent of radioisotopes

The story of rubisco goes back to the development of cyclotrons in the 1930s by Ernest Lawrence and Stanley Livingstone at the 'Rad Lab' (radiation laboratory) in Berkeley, California. A cyclotron accelerates charged particles, typically protons, into a high-energy beam that can be targeted onto some material of interest. The protons are injected into the centre of a circular vacuum chamber, and their flight is accelerated by rapidly reversing an electric voltage (thousands or millions of times every second) while applying an external magnetic field that bends their course into a circular path. With each acceleration the circle widens, so the protons spiral out from the source. The first 'proton merry-go-round' was just five inches in diameter, but larger diameters enabled more spirals, and so greater acceleration and higher energy. By the end of the 1930s, Lawrence had built a cyclotron with a sixty-inch diameter, which could produce a proton beam that spiralled up to nearly a fifth the speed of light before being directed onto the target material.[2]

By bombarding the nuclei of target atoms, a high-energy proton beam disrupts their atomic structure, producing isotopes – atoms with differing numbers of neutrons in the nucleus, giving each one a slightly different atomic weight. The chemical properties of an atom depend on the number of protons and electrons, but other properties such as radioactivity depend on the number of neutrons. As a rule, the number of protons and neutrons in the nucleus of an atom is roughly equal, and atoms that have more (or fewer) neutrons tend to be radioactive: the nucleus is unstable, meaning that its constituents rearrange themselves into more stable states over time, emitting energy or subatomic particles as they do so. Being hit by high-energy protons can dislodge neutrons or protons, or cobble on an additional proton, changing the configuration of the nucleus, frequently into an unstable state emitting radiation. While Lawrence was mainly interested in the nuclear physics, much of the early work with radioisotopes aimed to develop new medical applications, such as treating hypertension and arthritis as well as cancer (where radium was already being used).

But it was the fundamental applications in biology that made the biggest difference, with the advent of radioactive carbon isotopes, initially carbon-11 in the late 1930s. Carbon

---

2 This is sufficiently below the speed of light not to require correction for the effects of relativity. Because $E = mc^2$, as particles are accelerated closer to the speed of light their mass and behaviour in electromagnetic fields changes, which has to be taken into consideration in synchronising the reversals in electric field with the flight of charged particles. As a result, the higher-powered descendants of cyclotrons are called synchrotrons. The largest cyclic particle accelerator is the Large Hadron Collider at CERN, in Geneva, with a diameter of 27 km.

normally has an atomic weight of 12, with six protons and six neutrons in its nucleus. Bombarding crystals of boron – carbon's neighbour in the periodic table – with deuterons (heavy hydrogen nuclei, composed of a neutron as well as a proton) transmutes some of the boron into carbon-11, usually notated ^{11}C, as it picks up an additional proton. This conversion of one element into another was of course the goal of alchemists, who sought to turn lead into gold. When in 1901 Frederick Soddy realised that radioactive thorium was converting itself into radium, he cried out to Ernest Rutherford, 'Rutherford, this is transmutation!' Rutherford responded 'For Christ's sake don't call it transmutation. They'll have our head off as alchemists!' By the 1930s, though, transmutation was so commonplace that it had become a matter for respectable experimentation. Indeed, the theory was so unreliable that this experimentation was necessary. The results were often unpredictable, depending on the energy and composition of the beam (protons or deuterons, for example) as well as the duration of collisions and the target material. ^{11}C has a half-life of just 20 minutes, meaning that half of it will have decomposed back to boron (by emitting 'positrons') every 20 minutes. After two hours, nearly 99% of ^{11}C will have decayed back to boron, with its radioactivity accordingly declining to a hundredth of the starting point. Before much longer, the residual radioactivity will have dwindled to undetectable levels. That didn't give researchers long to do much with it, but it was a start. And it certainly added to the drama.

The pioneers of carbon isotopes were chemists caught between physics and biology. Their story is the stuff of scientific mythology, even if it is less widely known to the general public, and it is worth recounting not only because it explains our focus on sugars at the heart of metabolism, but also because it betrays how structural problems in politics can lead to structural problems in science – and how unhelpful scientific dogmas can take hold.

The 'Big Problem'

Enter stage left: Martin Kamen and Sam Ruben. This pair of isotope chemists joined Ernest Lawrence's Rad Lab in the mid-1930s. Kamen sets the scene in a charming autobiography. Lawrence himself and Robert Oppenheimer, affectionately known as Oppie (who goes on to lead the Manhattan Project) give a special seminar for Niels Bohr, the pioneer of quantum theory, who happens to be visiting the lab. They present exciting new results on the disintegrations of platinum when exposed to high-energy deuteron beams. Lawrence gives a short presentation of the data, before Oppie launches into a 'stupefyingly brilliant exposition of its theoretical consequences', leaving the audience 'in dumb admiration'. When Oppenheimer pauses, Bohr raises his hand diffidently and remarks in broken English that he has much difficulty in believing that the data could be valid, and therefore that the theory could have any real basis. The problem lies in the chemistry, not the physics. In the panic that ensues, Lawrence asks Kamen and Ruben to repeat the experiments. They show that Bohr is correct: the problem is chemical contamination from laboratory dust baked onto the platinum foil. Luckily, this embarrassing mishap doesn't stop Lawrence from receiving the Nobel prize for his pioneering work on cyclotrons in 1939.[3]

---

3 The very idea behind the cyclotron had been criticised by no less a figure than Albert Einstein, who compared the idea of firing subatomic particles at a target to be something akin to shooting birds in the dark, in a country where there aren't many birds. His point was that the atom is virtually entirely space,

This bonding experience cemented a strong friendship between Kamen and Ruben, but also made clear the importance of proper chemical analysis and preparation in a physics lab. For the first time, chemistry, even biology, was encouraged. That segued into the pair thinking about the possible applications of cleanly prepared radioisotopes such as ^{11}C in biology. Their initial plans to study sugar metabolism in rats were convoluted, to put it mildly. The idea was to incorporate ^{11}C into sugars by first exposing plants to CO_2 enriched in ^{11}C, which the plant would duly convert into glucose through photosynthesis. The sugar could then be isolated from the leaves, purified and fed to the rats. That was always a tall order, given the 20-minute half-life of ^{11}C, and not surprisingly ended in dejection. Hardly any of the ^{11}C ended up in glucose or related carbohydrates. It was hard to take, but it also opened up a window on other possibilities ... what if plants didn't make sugars such as glucose by photosynthesis at all? What if they were barking up the wrong tree?

Then Ruben had a brainwave. 'Why are we bothering with rats at all? Hell, with you and me together we could solve photosynthesis in no time!' The pair became obsessed with what Kamen called the 'Big Problem': finding the first product of CO_2 fixation in plants. If it wasn't glucose, then what was it? Earlier that year, Krebs had spelled out the individual steps in the breakdown of glucose to CO_2 and H_2O. Photosynthesis was supposed to do more or less the reverse, yet it remained a black box. What were the steps by which thin air was transformed into the substance of life, starlight into matter? This was the

---

the nucleus taking up a tiny fraction of the space of any atom (OK: about 0.000000000004% of the hydrogen atom) and therefore the total target of the beam. But he hadn't anticipated the sheer number of particles in a cyclotron beam – it was the blunderbuss to end them all.

'Big Problem', no less than the anatomy of a miracle, to be played out against the hubristic backdrop of the nuclear bomb, Man playing God.

Their idea was simple, even if the execution turned out to be fraught with difficulty. Feed the plant with CO_2 enriched in ^{11}C, give it a little time to fix the CO_2 to form new organic molecules incorporating the ^{11}C, then stop the biochemistry dead in its tracks by dropping a leaf into boiling alcohol. If the reactions of photosynthesis are stopped quickly, after a few minutes, then most of the ^{11}C should be incorporated into the earliest products of CO_2 fixation. Stopping the reactions after longer periods would give later products. All they had to do was isolate the radioactive molecules and use the standard methods of organic chemistry to identify them. By building up a time-course, Kamen and Ruben could then track the path of carbon in photosynthesis, giving a route map for the whole pathway.

Beautiful, simple … and doomed to failure. Problems dogged them from the beginning. Their venture was a sideline for the Rad Lab, so they were only granted occasional access to the powerful 37-inch cyclotron. When they did get access, it was all systems go. Kamen had to separate the radioactive CO_2 from the boron oxides that he had been bombarding with deuterons, no easy task in itself. The precious sample was then dashed across to the lab in the 'rat house', where Ruben and Zev Hassid were at the ready with their Geiger counters and all the reagents for analyses, which needed to be completed within about four hours, before the radioactivity had dimin-ished to undetectable levels. Kamen imagined how it must have looked to someone outside – 'three madmen hopping about in an insane asylum', often in the middle of the night. Even the analyses turned out to be hopelessly complex. The total dose of radioactive CO_2 was so low that they could not identify the trace amounts of new organic molecules formed. What little radioactivity they could detect was typically associated with larger proteins; the small organics they were seeking had bound tightly to the proteins' surface. They were, after all, physical chemists in a physics lab, not biochemists, and hardly trained for this line of work.

But their work was raising the curtain on a new era, and the photosynthesis community could see the raw potential despite all the setbacks. Kamen and Ruben made real progress (though Zev Hassid was obliged to quit for health reasons). The prevailing idea in the late 1930s was that CO_2 bound directly onto chlorophyll, the green photosynthetic pigment in plants. When activated by absorbing light, the chlorophyll was thought to transfer electrons to CO_2 to generate formaldehyde (CH_2O), which would then polymerise to form glucose ($C_6H_{12}O_6$), retaining the same proportions of carbon to hydrogen and oxygen. Ruben and Kamen showed that no radioactivity accumulated in formaldehyde (and for that matter virtually none in glucose). The incorporation of radioactive CO_2 did not even need light – it could take place in the dark.[4]

Even so, while successfully disproving these earlier ideas of photosynthesis, Kamen and Ruben could only make one positive step towards identifying the actual path of carbon fixation: they showed that the first product was a carboxylic acid, with the same kind of chemistry as Krebs-cycle intermediates. This could have signalled the centrality of carboxylic acids and the

---

4 This unexpected finding took biologists some time to accept, yet CO_2 fixation in the dark had already been shown to occur in autotrophic bacteria. Indeed, Krebs himself had made the bold claim (based on manometry) that even animal tissues could fix CO_2, but he was unable to prove it using ^{11}C isotopes, as he hoped, for the war had just begun and he couldn't make his planned trip to Harvard to test the finding with the cyclotron there.

Krebs cycle for CO_2 fixation and so all of core metabolism, but the path was twisted. I mentioned in Chapter 1 that the carboxylate group is essentially a CO_2 unit, and it can break off as such. That's true the other way around too; when CO_2 is added onto another organic molecule, you end up with a carboxylic acid. The first step is deceptively simple, for we'll see there's a lot hidden behind this façade:

$$RH + CO_2 \longrightarrow RCOOH$$

Note that the 'R' here refers to a (then) unidentified chemical group. Ruben got so far as suggesting that R was actually a sugar phosphate (correct as it turned out), but soon afterwards history took a tragic turn.

Slow disintegration

1940. The war had begun in Europe, but the United States had not yet entered the fray. In that year of loaded tension, Kamen and Ruben made their most momentous discovery of all – ironically, nothing to do with either the war or photosynthesis directly, but a discovery that was to transform research into the metabolic basis of life, to say nothing of archaeology and human history: the formation of the slowly disintegrating radioisotope of carbon, ^{14}C, with its half-life of 5,700 years.

In many respects, ^{14}C was more of an invention than a discovery, although it later turned out to exist naturally, albeit at very low abundance. The first hints that a relatively stable radioisotope of carbon existed came from cloud chamber work with nitrogen gas. A cloud chamber is a mesmerising device, perhaps the closest anyone can come to actually 'seeing' subatomic particles with the naked eye. Suddenly expanding the volume of the chamber (in the early days by using a piston as the base of the chamber) cools the gases within, making them more likely to form clouds of vapour. As in the more familiar clouds in the sky, where water droplets condense around dust particles known as nucleators, the clouds in a cloud chamber form around tiny particles. The fleeting passage of a charged subatomic particle through a cloud chamber ionises molecules in its immediate vicinity (stealing electrons from them) which act as nucleators for cloud formation, leaving behind a vapour trail similar to that formed by an aeroplane. Here's the lovely part. The vapour trail formed depends on what kind of particle is passing through the chamber. A proton is tiny and fast moving. It whizzes through the cloud chamber, leaving a long, spindly vapour trail that vanishes in an instant, a ghostly apparition. Larger particles such as alpha particles (the nuclei of helium atoms with two protons and two neutrons) form shorter, stubbier trails.

What Martin Kamen noticed was that when he bombarded a cloud chamber containing nitrogen gas with neutrons emanating from a cyclotron,[5] he saw long thin vapour trails

---

[5] Neutrons? Yes, you heard right. Neutrons are produced from a high-energy beam of protons or deuterons hitting almost any target. They are also produced directly from the cyclotron itself, as particles are accelerated up to higher speeds and energies. At energies above ~10 million electron volts, deuterons begin to break down and release neutrons as well as gamma radiation (high-energy photons). The electricity demands to accelerate deuterons up to these speeds periodically caused power failures across the city of Berkeley in the 1930s. Only discovered in 1932 by Chadwick, the neutrons produced by cyclotrons were being used for cancer therapy by 1938, although the clumsy early work suggested that the side effects outweighed the benefits. Lawrence had successfully treated his own mother for cancer at that time, using one of his own machines; luckily, he used X-rays rather than neutrons, and his mother eventually outlived Lawrence himself (who died at the age of fifty-seven from ulcerative colitis, probably exacerbated by stress).

corresponding to protons, plus some short, stubby trails that had to be formed by much heavier charged particles. As nitrogen normally has an atomic mass of 14 (made up of seven protons and seven neutrons), hitting its nucleus with a flying neutron displaces a proton, leaving its distinctive vapour trail. What is left behind is a nucleus that now has eight neutrons and six protons. This atom still has a mass of 14, but the loss of one proton transmutes the nitrogen into carbon, with its atomic number of six corresponding to six protons. It took some persistence from Kamen, as the physicists around him, notably Oppenheimer, used the rudimentary atomic physics of the time to calculate that the postulated transmutation could not be taking place. But Kamen was a chemist and believed the evidence of his eyes. He knew it could only be ^{14}C and the theory had to be wrong (again). Kamen tried all kinds of ways of making ^{14}C on a larger scale and Lawrence for once gave him as much time as he needed on the largest cyclotrons to do so. Eventually he succeeded. By bombarding ammonium nitrate with neutrons spinning off from the cyclotron, Kamen succeeded beyond his wildest dreams. Working with Ruben to isolate radioactive CO_2 from potentially explosive ammonium nitrate sludge (which Lawrence later banned) they made so much ^{14}C they paralysed the Geiger counter.

It had been far from straightforward. For months, Kamen had been bombarding boron with deuterons, then ^{13}C-enriched graphite, eventually succeeding in producing a small amount of ^{14}C. After working around the clock for weeks, Kamen finally finished in the middle of the night, and left his sample on Ruben's desk before making his way home through a storm. There had been a murder that night across town and the police picked up a clearly deranged lunatic lurching through the rain, who turned out to be Kamen. He was only released when a hysterical witness to the murder showed no sign of recognising him. But the incident set the tone. The bombing of the US fleet at Pearl Harbor in 1941 was greeted with disbelief that swiftly morphed into fear, as San Francisco seemed only too real a target. The US government took over the cyclotrons at the Rad Lab to focus on the production of radioactive isotopes, notably uranium and plutonium. Kamen was put in charge of developing the new protocols. Ruben was set to work in defending the US coastline from invasions using the poison gas phosgene. Exhausted from working long hours so he could return to his beloved work on photosynthesis, Ruben fell asleep at the wheel and crashed his car. Nobody was seriously injured, but Ruben broke his arm. Soon afterwards, while preparing a batch of liquid phosgene in the lab, one arm in a sling, a cracked glass tube exploded in some fiercely boiling liquid air, spraying Ruben's jumper with the deadly gas. Knowing he may have exceeded the lethal dose, Ruben calmly carried the broken glassware outside to save others. He died the following day, not yet thirty years old, his lungs swelling up with fluids, his young family bereft. He'd never found the time to fill out the forms that would make them eligible for a federal pension.

Kamen was forced to abandon his work on photosynthesis too. An accomplished viola player, he was close friends with the virtuoso violinist Isaac Stern. Kamen had unsuspectingly gone along to a party hosted by the violinist that was also attended by some Russian attachés, with whom he remained briefly in touch. But Kamen had already drawn attention to himself for deducing from radiation energies that the US government must be running an atomic pile at Oak Ridge. This apparent treason – how could he possibly have known without an informant? – was the final straw. He was abruptly relieved of his position in the Rad Lab. His wife left him. Under surveillance from army intelligence, Kamen was prevented from taking any scientific position, and eventually found employment in the shipyards of San Francisco. After the war, the episode returned to mire him: Kamen was hauled before the House Un-American Activities Committee (as was Oppenheimer and many others). He was smeared by the press as 'Atomic Scientist Fired from Army Project after Talking to Reds', before finally clearing his name more than a decade later. While he did successfully return to science, Kamen never resumed his pioneering work on the path of carbon in photosynthesis, and never put his discovery of ^{14}C to use. That path was mapped out by others. Had Ruben lived, it seems a fair bet that the pair of them would have been awarded the Nobel prize for their discovery of ^{14}C. That didn't happen because the prize cannot be awarded posthumously. Instead, the discovery, coupled with the untimely demise of their partnership, set in train a peculiar history that still influences how we understand photosynthesis today, and especially central carbon metabolism – the notion that sugars are the backbone of biochemistry. Only the discerning power of ^{14}C could mark out the path by which carbon was fixed in photosynthesis at that time, and only the Rad Lab had enough ^{14}C to do the job. The dice were loaded.

The one true path

Keen to progress the work on photosynthesis, Lawrence hired Melvin Calvin, a colleague from the Manhattan Project, immediately after the war. The story has it that on the day of the Japanese surrender Lawrence told Calvin that 'Now is the time to do something useful with radioactive carbon.'

Calvin was by all accounts brilliant and inspirational, with a remarkable memory, but he was more interested in the photochemical wizardry of the 'light reactions' in photosynthesis, which he never cracked. Calvin took on the carbohydrate chemist Andrew Benson, to focus on what Calvin saw as the less exciting path of carbon in photosynthesis, ironically the work now inextricably linked with Calvin's name. Benson not only provided continuity, having been friends with Sam Ruben, but as it happened, had been entrusted with the world's entire supply of ^{14}C by Ruben, shortly before his death. Where Calvin was bubbling over with ideas, some of which were borderline crackpot (and I suspect that Benson was one of the few willing to tell him so), Benson was not given to highfalutin notions, but had an exceptional facility for thinking up clever experimental designs. Their progress over the next few years owed much to his ingenuity.

Calvin and Benson decided to work on the alga *Chlorella*, which could photosynthesise much faster than land plants at high light intensity. This was facilitated by Benson's famous 'lollipops', flattened glass flasks that could be illuminated from both sides, with a tap in the base to allow samples to be run off into boiling alcohol, killing the cells for analysis. Perhaps Benson's most decisive change was to adopt a new method of analysis that is familiar in its simplest form to all schoolchildren – paper chromatography. Benson made two clever changes to this method. He selected better solvents that would separate small, electrically charged molecules such as carboxylic acids and sugars in two dimensions (with the second solvent running perpendicular to the first). The second clever trick was to place a photographic film over the paper, which darkened in the presence of radioactivity. If the ^{14}C was incorporated into a specific type of molecule, this would migrate to a precise place on the paper, depending on its chemical properties. The radioactivity emanating from the ^{14}C would then betray its presence as a dark smudge on the photographic film. For example, a specific sugar that incorporated ^{14}C would always migrate to the same place on the paper, giving a diagnostic dark spot on the photographic film – an 'autoradiograph'. The bit of paper responsible for the dark spot could then be cut out, the sugar washed out, and its identity established by standard chemical analyses. The method was derided by some as 'spots on paper' (a very literal description) but it transformed the level of analysis possible.

With this experimental setup, stopping photosynthesis after as little as one minute already produced as many as fifteen spots, giving plenty of scope for confusion. Killing the algae after just ten seconds gave a much cleaner signal – just one spot, meaning that only one product gave rise to all the radioactivity; it had incorporated all the ^{14}C. Chemical analysis showed this to be a three-carbon carboxylic acid called phosphoglycerate:

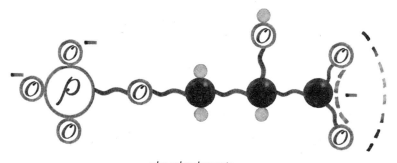

phosphoglycerate

Notice that the newly fixed CO_2 (highlighted with the dotted line) seems to be joined onto a two-carbon acceptor molecule, with a phosphate group also attached (PO_4^{2-}), as anticipated by Ruben and Kamen. It's hard to convey the excitement that one dark smudge signifying this unprepossessing little molecule must have caused. Think about it for a moment: what is more mysterious than the everyday miracle of photosynthesis, in which plants turn thin air into wood and leaves, flowers and fruit? How on earth do they do that? Here was the first tangible clue, which transformed nature's deepest magic into the language of science, of human comprehension, of the rational mind – a simple molecule that made sense. All that in one dark smudge. This is why scientists get so excited about the seemingly abstruse. That is no smudge, it is a whole world!

But it was also the only cheap insight. In less than a minute, the radioactivity had spread into many other products, each one of which took days, weeks, sometimes months, in the lab to identify, with the most careful chemistry. Amongst the most common radioactive spots were various (but not all) Krebs-cycle intermediates, notably pyruvate (C3), succinate (C4) and malate (C4), as well as a few amino acids and some sugars. What did these say about the path of carbon? Following the logic of Krebs himself, Calvin and Benson were seeking a cycle: a C2 acceptor picked up a CO_2, giving a C3 acid. If cells were not to run out of C2 acceptors they would need to regenerate them, or everything would grind to a halt – hence the cycle. They knew that later products included at least two C4 carboxylic acids. So presumably there was another step, in which a second CO_2 (plus more hydrogen) was added on to give a C4 acid – malate or succinate. That could then be split in two, regenerating the C2 acceptor plus another C2 molecule. This brand new C2 molecule could now be siphoned off to make sugars or amino acids, eventually proteins and DNA. All that made perfect sense, except that ... none of it was true. Starting in 1948 with a paper in *Science* entitled 'The Path of Carbon in Photosynthesis', Calvin, Benson and their colleagues published a stream of papers, each one with the same title numbered with a Roman numeral. By 1952, they had got up to number XX, though frankly they weren't much closer to the right answer.[6]

The problem was that they were befuddled by the Krebs-cycle intermediates which kept on cropping up. In the end, it turned out that *none* of these lay on the one true path of photosynthesis as we know it. Those Krebs-cycle intermediates appeared so often because they are so central to all metabolism – a salient point that we'll return to in a while. But here they were just misleading.

The key turned out to be Benson's personal discovery of a five-carbon sugar-phosphate called ribulose bisphosphate (found while Calvin was away, recuperating from a heart attack that he'd had during a budget meeting[7]). This sugar accumulated when the algae were starved of CO_2 – exactly what a CO_2 acceptor should do! So the acceptor was not a C2 molecule at all, but a C5 molecule that split in half when CO_2 was added on, giving two C3 phosphoglycerate molecules. The rest was mostly arithmetic, albeit a strange numbers game. Three complete turns of the cycle produce six C3 molecules. One of

---

6 It is worth reflecting that paper after paper published in top journals can be essentially wrong. Please don't think that these papers were especially bad. Far from it. Science is hard and we are all wrong a lot. The harder the problem, the more wrongness there will be. If you want to take a moral away from this, then assume that most published papers are at least partly wrong. That helps to explain why scientists squabble a lot in public. But these arguments are part of the way in which science corrects itself. Publishing detailed papers lays out exactly where scientists are wrong, and how to ask better questions. Unlike any other human endeavour, to my knowledge, the scientific method is a ratchet for improving answers over time. Being wrong is part of the ratchet, just as detrimental mutations are part of natural selection, driving the evolution of life's wonders.

7 Calvin's wife Genevieve took matters in hand: under her strict dietary regime, Calvin lost 30 kg in weight and quit smoking. It would be unfair to blame the budget meeting for anything but the timing of his heart attack.

these could be siphoned off to be used anywhere in metabolism, while the other five C3 molecules went through a rejigging process to regenerate three C5 molecules – all the ribulose bisphosphate needed to enable another three turns of the cycle. The full scheme, now in textbooks as the Calvin–Benson cycle, was published in the classic 1954 paper, 'The Path of Carbon in Photosynthesis. XXI.' Here's a slightly modified scheme from that paper, which captures the essence of the cycle:

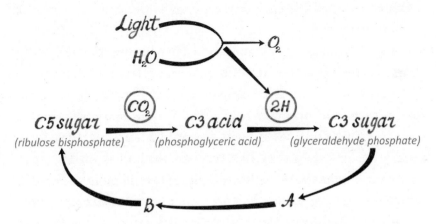

The details needn't concern us. Suffice to say this is sugar metabolism. Phosphoglycerate is fed hydrogen, derived from the solar-powered splitting of water (which we'll come to in Chapter 4) to form a three-carbon sugar called glyceraldehyde phosphate. That is converted, by way of other sugary intermediates (let's just follow Calvin and Benson and call them A and B here[8]) to regenerate the five-carbon acceptor molecule, the sugar ribulose bisphosphate.

---

8 Did you really want to know what they are? A is fructose phosphate, a six-carbon sugar. B represents several different sugars, including xylulose phosphate (C5), erythrose phosphate (C4) and sedoheptulose phosphate (C7). Yes, it's absurdly complicated. The numbers game works roughly like this. The C6 is split into a C2 plus a C4. The C2 is joined onto a C3 to make a C5. The C4

Does that name ring a bell? Remember the enzyme rubisco, the most abundant breakfast cereal in the world, which technically stands for ribulose bisphosphate carboxylase-oxygenase. What we see here is the carboxylation of ribulose bisphosphate (meaning that CO_2 is added onto it) to form two molecules of phosphoglycerate. The 'oxygenase' activity occurs when O_2 replaces CO_2 in the figure, to make other products, including ultimately some carboxylic acids that confounded so many of Calvin and Benson's early attempts to understand the path of carbon fixation.

That seminal 1954 paper was the culmination of more than just the cycle. For reasons that were never made explicit, Calvin abruptly dismissed Benson with the words 'It is time to go.' Benson had no place to go, and Calvin made no obvious effort on his behalf. What went wrong? Calvin's own cherished ideas on photochemistry had recently imploded, so his greatest achievements in the field were more attributable to Benson, which must have rankled. And I suspect that Benson didn't mince his words in telling Calvin what he thought of some of Calvin's more wayward ideas. Benson was certainly capable of being caustic in his judgements. Calvin's wife Genevieve told Benson that Calvin would have another heart attack if he stayed. I wonder if Benson treated Calvin with some measure of disdain; he had certainly stopped telling him what he was doing on a daily basis. Be that as it may, Calvin only diminished

---

is joined onto a C3 to make a C7. The C7 is split to make a C5 plus a C2, and the C2 is joined to a C3 to make a C5. So now we have three C5 molecules, but none of them is the correct C5; all of them need to be converted through more biochemical shenanigans into the C5 acceptor, ribulose bisphosphate. Some of this complex chemistry requires energy in the form of ATP, which is provided by photosynthesis too. We'll come to how that works later. If your head is beginning to spin, just remember that all this sugar chemistry is what I said deceived the whole field for generations, so don't worry about it.

himself through his later actions. Seven years after Benson had gone, in 1961, Calvin was awarded the Nobel prize, alone, for 'his research on carbon dioxide assimilation in plants'. In his Nobel lecture, Calvin mentioned Benson just once, in passing. Even more disturbingly, in 1991 Calvin published his auto-biography, in which he airbrushed Benson from history. In a 175-page book there is no mention of Benson – not a word, or a photograph (in a book with 51 pictures), or even a citation of any co-authored paper, despite an extensive bibliography. He didn't even cite the classic 1954 paper that made Calvin's name synonymous with the 'Calvin cycle'. As Benson reflected many years later, Calvin 'didn't have to do that. He could have done it right.'

I say the 'Calvin cycle'. For many years that was indeed the established textbook name, and it certainly trips off the tongue – ease is some sort of virtue. I am guilty here of not men-tioning another key player in this story, James Bassham, who was instrumental in working out exactly which carbon atoms picked up a radioactive tag, and precisely when – on a timescale of seconds. Many now refer to the Calvin–Benson–Bassham cycle, which is fair, if admittedly a bit of a mouthful.[9]

---

9 Science is almost always a team effort, frequently with large groups of people, all of whom contribute in important ways, albeit to varying degrees, just as large teams do in making movies. To what extent does one individual's stamp make a movie or a scientific discovery? We can recognise the creative vision of a director, editor, writer or producer in films. To think of a movie only as an ensemble piece created by a tight-knit team of specialists would be to miss something important from Hitchcock's or Sergio Leone's films. But to imagine they could have made the movie alone is absurd. Likewise, it's fair to recognise the drive, creativity, vision or shear tenacity of individuals in science. I'm singing the praises of a few in this book, with their warts and all. It is fundamentally misleading to see science as the colossal achievements of indi-viduals, even if standing on the shoulders of giants; but genius and transfor-mative vision do exist. Hubristic scientists will always claim priority or genius.

My point here is not about fairness but personal myth. Calvin was an unusually persuasive and charismatic man. He even featured on a 2011 series of US postage stamps dedicated to great American scientists, alongside Richard Feynman, Barbara McClintock, Linus Pauling and Edwin Hubble. Calvin's claim to fame was the one true path of carbon in photosynthesis – this was *the* pathway of autotrophy, and Calvin was king. It was all about sugars. Here's a typical assertion from the 1970s: 'One important property shared by all autotrophic species is the assimilation of CO_2 via the Calvin cycle.' You can imagine that if anyone had the temerity to claim another way of fixing CO_2, that amounted an insurrection against the unity of bio-chemistry, a rebellion against the king. That's exactly what happened next – and why it took so long to make any headway. Has it really made headway, even today? The idea that CO_2 fixation makes sugars is still the paradigm that most students imbibe today – the central dogma of metabolism. And so long as it stays that way, we will never understand the deeper currents in the chemistry of life.

The reverse Krebs cycle

Berkeley casts a long shadow over the photosynthesis world. Five years after Calvin's Nobel lecture, another luminary of photosynthesis research, also at Berkeley, published a paper that might as well have come from a parallel universe. Daniel Arnon was Polish. He was drawn to Berkeley in the 1930s by the California novels of his hero Jack London. He remained at Berkeley his whole career, ironically garnering a reputation as a

---

Ecclesiastes was close to the bone when he said that all is vanity and vexation of spirit.

'European' professor, meaning somewhat authoritarian in discipline, but perhaps also reflecting his eloquent, argumentative and philosophical approach to science. He loved to play devil's advocate, even in his own lab meetings. Not surprisingly, Arnon did not get along with Calvin. In one abortive attempt to reconcile the two groups a joint seminar descended into mayhem. About ten minutes into the seminar, Calvin and Arnon started arguing about reaction rates and their consequences for the abundance of ^{14}C-labelled intermediates. The speaker fell silent as the pair continued to argue, 'Calvin with the precision of a physical chemist and Arnon with the reasoning of a philosopher.' Calvin later made it a point of honour never to recognise members of Arnon's group. These altercations obliged visitors to Berkeley to conceal any arrangements to meet the rival faction, lest word got out that they had betrayed their host.

With fitting symmetry, the discovery that made Arnon's name was published in 1954, the same year as the Calvin–Benson cycle – and even in the same journal. Arnon had shown that photosynthesis does more than make new organic molecules: it also provides the energy that's needed to power their synthesis by 'photo-phosphorylation' – ATP synthesis powered by light. This process turned out to work in basically the same way as respiration. Light powers the splitting of water, stripping the 2H from H_2O through the auspices of chlorophyll. The 2H in turn splits into electrons and protons, which then follow different paths, just as they do in respiration. Remember Peter Mitchell's 'proton-motive force' from Chapter 1. In the case of photosynthesis, a current of electrons, streaming from water, powers the extrusion of protons across a membrane, generating a proton-motive force. This drives the synthesis of ATP through that extraordinary nanoturbine, the ATP synthase. Plants use almost exactly the same machinery as respiration, except that it is buried deep within the chloroplasts. Arnon showed that these 'light reactions' of photosynthesis generate ATP as well as 2H, while Calvin and Benson showed that the 2H are used to make the three-carbon sugar glyceraldehyde phosphate, through reactions that can run in the dark as well as the light.

Given that background, picture the reception for Arnon's 1966 paper. Working with Mike Evans and Bob Buchanan, Arnon showed that the Calvin–Benson cycle was not the only pathway of CO_2 fixation after all. In fact, they made it look decidedly second rate. Because one turn of the Calvin–Benson cycle incorporates just one molecule of CO_2, they wrote, 'One complete turn gives, on balance, a net synthesis of 1/3 molecule of triose phosphate.' In contrast, they introduced the green sulfur bacterium *Chlorobium thiosulfatophilum*, which lives by photosynthesis in stinking, sulfurous waters, such as hot springs. This bacterium, they reported, reverses the Krebs cycle. In doing so, it 'incorporates four molecules of CO_2 and results in the net synthesis of oxaloacetate, a four-carbon dicarboxylic acid, which is itself an intermediate in the cycle. Thus, beginning with one molecule of oxaloacetate, one complete turn of the cycle … will regenerate it and yield, in addition, a second molecule of oxaloacetate formed by the reductive fixation of four molecules of CO_2.'

This was turbocharged carbon fixation! More than that, this is *autocatalysis*, capable of giving rise to an exponential lift-off. If one complete turn of the cycle generates two molecules, then two turns would make four, three turns make eight, four turns sixteen, and so on … It could drive growth like a … well, like a bacterium. Here's what they had in mind (I'm just slightly simplifying their original diagram to highlight the key points):

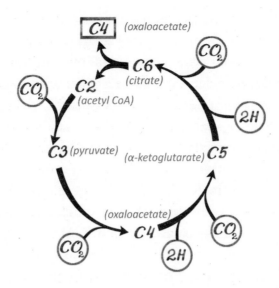

Instead of pulling out hydrogen and CO_2 from organic molecules to generate ATP, the reverse Krebs cycle does the opposite:
it uses CO_2 and hydrogen to make new organic molecules (and
costs some ATP). One complete turn of the cycle regenerates
the starting point plus another molecule – oxaloacetate in the
figure above. Because oxaloacetate is itself an intermediate, it
too can spin round the cycle to generate any of the other intermediates. In principle, any intermediate between C2 and C6
could be bled off and used for biosynthesis. As we've seen, these
carbon skeletons are the currency of biochemistry. Do you want
to make an amino acid? Your starting point is right here in
the Krebs cycle. Do you want to make fatty acids or isoprenes
for cell membranes? You start with acetyl CoA. How about
sugars? Start with pyruvate. Nucleotides for RNA or DNA?
Your building blocks are amino acids (derived from the Krebs
cycle) and sugars (ditto). Please note that I'm not showing the
full cycle here – that's shown on page 287. But you can see
here that the reverse cycle is slightly longer than the standard
cycle running backwards: it includes several extra carboxylic

acids that are not usually thought of as Krebs-cycle intermediates, notably acetyl CoA and pyruvate (plus that metabolic link between carboxylic acids and sugars, phosphoenol-pyruvate, as you can see on page 287). The reverse cycle integrates them into a biosynthetic engine at the very heart of metabolism, in a way that makes perfect sense. The Krebs cycle provides C2 to C6 carbon skeletons, the Lego bricks of life, to make virtually everything else.

The idea that the Krebs cycle running in reverse could drive growth through carbon fixation was not new; it had been suggested in the late 1930s, and certainly influenced Ruben and Kamen's thinking. The ubiquitous appearance of Krebs-cycle intermediates in the autoradiographs of Calvin and Benson also pointed to their importance, but actively misled them for half a decade – it turned out that none of these intermediates were part of the Calvin–Benson cycle. So when Arnon and his colleagues used Benson's own methods of ^{14}C autoradiography to report even more Krebs-cycle intermediates, it must have felt as if they were reintroducing mistakes from the past. There's a story of a dinner in Trondheim in 1980 when Calvin sat next to the microbiologist Reidun Sirevåg. He asked her what she was working on. 'I hesitate to say this', she replied, 'I am working with an autotrophic, photosynthetic bacterium, which fixes CO_2 but not by the Calvin cycle.' 'I don't believe it', Calvin retorted with a grin, ending the conversation.

It's no coincidence that the 1966 paper was published in *Proceedings of the National Academy of Sciences of the USA*, of which Arnon was a member. That honour meant he could publish in the Academy's journal without the normal formality of peer review, which might well have killed it in embryo.[10]

---

10 Peer review is often lionised as the 'gold standard', and at its best can indeed

In the end, the wider field did not come to accept the reverse Krebs cycle until the 1980s (a quarter of a century later) when gene sequences corroborated that *Chlorobium* really doesn't possess the Calvin–Benson cycle, but it does have all the requisite genes for the reverse Krebs cycle. Just how difficult it was for these notions to take root is perhaps best illustrated by the fact that Krebs himself never discussed the reverse cycle, even in his short but influential paper on the evolution of metabolic pathways in 1981. By that time, there had been several decades of the dogma that photosynthesis is all about sugars. And of course in plants, photosynthesis is indeed about sugars. But there's a telling irony to that statement. The fact that Krebs-cycle intermediates don't contribute to carbon fixation in plants meant that their importance in autotrophic metabolism became obscured, as the focus shifted elsewhere. I was astonished when I first looked at Benson's autoradiographs: I had expected to see sugar phosphates but instead I saw Krebs-cycle intermediates everywhere – the same carboxylic acids that threw Calvin and Benson off the trail for half a decade. In one

---

work extremely well. But it is also intrinsically conservative, as long argued by Don Braben (I would strongly recommend his books, notably *Scientific Freedom*). The problem is twofold. First, peers are frequently antagonistic towards each other, as they are by nature competitors (or, just as bad, collaborators) for fame and fortune (papers and grants), all of us being human. Second, any revolutionary ideas, by definition, overturn the applecart, meaning the lifetime achievements of peers are diminished, even disproved. The greatest of scientists – the greatest human beings – will put their own emotions aside, stand on a soapbox, and proclaim their errors and misinterpretations. But it's normally less clear-cut than that, and as we've seen here, scientists often fail to rise above our own baser instincts. For these reasons, Braben argues that we need to develop better ways of assessing radically new ideas in science – the kind that lead to fundamentally new ways of seeing the world, and which did so much to revolutionise our understanding of life, the universe and everything in the twentieth century.

sense these were not misleading at all. They pointed to something equally important: sugars are peripheral. They're needed for the Calvin–Benson cycle (and of course to make nucleotides) but for everything else, materials need to be shunted off to the true metabolic hub of the cell, the Krebs cycle. Indeed, the fact that the Calvin–Benson cycle is so peripheral to the rest of metabolism actually makes it easier to incorporate as an additional metabolic unit, as it can be regulated independently from the core pathways of central metabolism – the cycle can be switched on or off if conditions change without conflicting with other essential cellular processes.

In any case, *Chlorobium* restored the Krebs cycle to its rightful place, at the metabolic heart of the cell. I admit that to most people, anaerobic sulfur bacteria may seem a little esoteric, even if they do organise their metabolism in an exemplary way. But that's exactly the problem. They will never replace plants in our affections. They should in our understanding. These bacteria offer a far more coherent view of evolution and its implications for our own health. So please, when you think of photosynthesis, don't just think of my clichéd images at the beginning of this chapter – spare a thought for green sulfur bacteria with their illuminating metabolism.

How to power growth

The problem facing Arnon and Buchanan was not only the unity of biochemistry and the dominating personality of Calvin; the idea of a reverse cycle also seemed to go against thermodynamics. Szent-Györgyi and Krebs knew well that parts of the Krebs cycle could run backwards, even in animal cells, although the reasons for that were opaque. Krebs had even posited the carboxylation of pyruvate to oxaloacetate in

1939 (the C3 to C4 step in the scheme on page 101), which was confirmed in 1940. In 1945 the great Spanish biochemist Severo Ochoa, working in New York, showed that animal cells could carboxylate α-ketoglutarate to make citrate (the C5 to C6 step above). This radical finding was hard to comprehend, as it still is even today. Plants were supposed to fix CO_2, not animals, yet here were animal tissues – our own tissues! – behaving as if they were plants. Even so, the remaining two carboxylation steps (C2 to C3 and C4 to C5) were taken to be biochemically impossible, dismissed as energetically too unfavourable. That meant the cycle as a whole was deemed to be irreversible.

Yet tiny sulfur bacteria could apparently reverse the Krebs cycle without problem. They did so by using a small, red protein discovered in the early 1960s which seemed to be capable of magic – this same protein was essential for photosynthesis to make ATP *and* to fix CO_2. Now it seemed that it could spin the Krebs cycle backwards too, driving growth. This marvellous, mysterious adaptor was named ferredoxin.

Ferredoxin is red because it contains iron. Specifically, it binds to itself one or two tiny mineral lattices, each one just a few atoms in size, known as iron–sulfur clusters. When bound to ferredoxin, these clusters have a potent ability to transfer electrons. It's not easy to force electrons onto ferredoxin – that's where the sun comes in. By exciting chlorophyll to steal electrons from water (or any other donor such as hydrogen sulfide) light drives the current of electrons in the photosynthetic membranes, powering ATP synthesis by photophosphorylation. At the end of the line is the terminal electron acceptor ferredoxin. Ferredoxin packs a punch. It forces its electrons onto CO_2 (indirectly in the Calvin–Benson cycle) or onto those two most recalcitrant Krebs-cycle intermediates – acetate (C2) and succinate (C4) – enabling the Krebs cycle to run backwards right the way round, to fix CO_2. Remember my portraits of succinate and acetate in Chapter 1 (pages 46 and 57) with their stout bellies and Cheshire Cat grins ... they're far too placid to react unless poked.

The basic chemistry is similar for both acetate and succinate, requiring CO_2 to be joined onto a virtually inert carboxylate group. Remember that the carboxylate group has a structure of -CO_2, so the CO_2 has to be bound onto something very like itself. Ferredoxin (Fd) takes the electrons from the 2H and transfers them onto CO_2, enabling it to form a carbon-to-carbon bond:

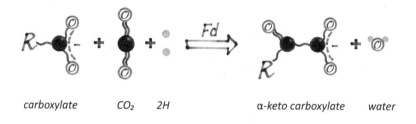

carboxylate CO_2 2H α-keto carboxylate water

Let's not worry about exactly how that works here (it requires several steps; see Appendix 1 if you want to know – the process is rather beautiful). The critical point here is that ferredoxin has a biologically unparalleled ability to press electrons onto even the most unreactive molecules. But that comes at a cost. Ferredoxin reacts spontaneously with oxygen, becoming readily oxidised by even low levels of the gas. So in the presence of oxygen the reverse Krebs cycle usually grinds to a halt. It's even worse than that, because when oxygen picks up single electrons from ferredoxin it is itself transformed into a reactive 'free radical' (a molecule with one or more unpaired electrons, typically making it more reactive). Notoriously, oxygen free radicals can go on the rampage, initiating lengthy chain reactions that oxidise lipids in cell membranes, inactivate proteins and mutate

DNA, causing disproportionate damage. In short: bad news. We'll say more about free radicals in Chapter 6, but here let's just acknowledge that they are real and can cause trouble.

That fact explains a lot. We'll look into how rising oxygen levels play havoc with the reverse Krebs cycle later. For now, let's note that the bacteria that use it today are normally restricted to environments with very low oxygen levels. Their sensitivity to oxygen probably explains why the Calvin–Benson cycle rose to dominance in cyanobacteria and plants, which produce oxygen as a waste product of photosynthesis. Plants still depend on ferredoxin, but keep its levels to a strict minimum, transferring its electrons straight to $NADP^+$ (nicotinamide adenine dinucleotide phosphate) to form NADPH like this:

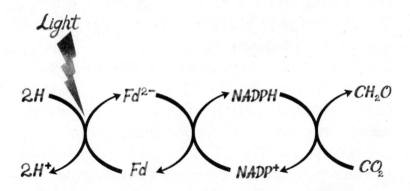

Transfer of electrons from 2H to CO_2 to form organics, $C(H_2O)$, in photosynthesis, powered by sunlight. The 2H could come from H_2S, H_2 or H_2O depending on the type of photosynthesis – but all need sunlight to transfer the electrons from 2H onto ferredoxin (Fd) to form Fd^{2-}. In the presence of oxygen, Fd^{2-} is dangerous as it can generate reactive free radicals. This eventuality is largely avoided by Fd^{2-} swiftly passing its electrons on to $NADP^+$ to form NADPH, rather than directly to CO_2.

Rather than dealing in single electrons, like ferredoxin, NADPH transfers pairs of electrons, which makes it far less reactive with oxygen, while retaining just enough power to push its electrons onto other molecules.[11] NADPH doesn't have the potency to drive the Krebs cycle backwards, but it can promote the sugar chemistry of the Calvin–Benson cycle. As so often happens in evolution, this sprawling cycle seems to have been pieced together on the hoof by splicing two existing pathways that produce sugars from Krebs-cycle intermediates (gluconeogenesis and the pentose phosphate pathway, if you wondered). The only thing missing was our friend, the enzyme rubisco. Amazingly, rubisco now turns out to be widespread in ancient bacteria, doing a totally different job: degrading sugars derived from the RNA of other cells, to support heterotrophic growth (growth fuelled by eating other cells). So the Calvin–Benson cycle is a Frankenstein's monster of a cycle, which just happens to work, albeit not very well; recall the excruciatingly slow turnover rate of rubisco and consider how few CO_2 molecules are fixed by a single turn of the cycle. It was a quick fix cobbled together in difficult circumstances. And it stuck because it worked.

The need to prevent ferredoxin reacting with oxygen might also explain the propensity of rubisco to fix O_2 through the

---

11 $NADP^+$ is a phosphorylated form of NAD^+ (nicotinamide adenine dinucleotide), which we met in the previous chapter as the main carrier responsible for transferring 2H from Krebs-cycle intermediates to oxygen via the respiratory chain. As a rule of thumb, NADH transfers electrons in breakdown reactions linked to respiration, whereas NADPH is used for the synthesis of new molecules. While there is little difference in their chemistry, the two carriers are maintained at very different ratios, enabling them to promote distinct types of reaction. $NADP^+$ is normally present as NADPH, the form that is replete with electrons, whereas NADH tends to offload its electrons, leaving it mostly in the NAD^+ form. We'll come back to the significance of NADPH in Chapter 5.

apparently futile process of photorespiration. Very little in evolution is genuinely futile; if it survives natural selection there is usually a reason. In the case of rubisco, think what happens if CO_2 levels fall while O_2 levels rise inside a leaf (when the stomatal pores are closed). Now rubisco is obliged to slow down because its substrate, CO_2, is in short supply. This means that NADPH cannot pass on its electrons to regenerate $NADP^+$. As a result, ferredoxin in turn is unable to pass on its electrons, and so it becomes reactive with oxygen, just when oxygen levels are rising. To stave off catastrophe, rubisco consumes oxygen instead. Photorespiration converts NADPH back to $NADP^+$, allowing ferredoxin to offload its electrons again. So it could be that photorespiration acts as a safety valve, lowering the levels of reactive ferredoxin and oxygen simultaneously, staving off an impending catastrophe. Certainly, this can cost plants a sizeable yield, but at least they survive to die another day. The penalty for replacing rubisco with an 'improved' enzyme could be sudden death. No doubt we'll find out in due course.

All that reasoning is consistent with a relatively late evolution of the Calvin–Benson cycle in a world of rising oxygen. The implication is that the reverse Krebs cycle was more widespread on the early Earth before the rise of oxygen. It has even been linked to the origin of life itself, as we'll see in Chapter 3. That view makes good sense, insofar as the reverse Krebs cycle is central to metabolism. Presumably, the reverse cycle preceded our own flipped version, explaining why the Krebs cycle is still the biosynthetic hub of the cell. Yet until recently, its distribution across the tree of life seemed disappointingly patchy, even among those anaerobic bacteria that shun oxygen. Since its discovery, several other pathways of carbon fixation have been discovered – now there are a grand total of six known pathways. None of these is as elegant as the reverse Krebs cycle, although they do all involve carboxylic acids, making them all more central to the structure of metabolism as we know it than the wonky Calvin–Benson cycle.

Two recent discoveries hint at the antiquity of the reverse Krebs cycle. The first relates to the need for a supply of ATP and ferredoxin for the cycle to spin at all. As a rule, these derive from photosynthesis. None of the most ancient bacteria (or their cousins, known as archaea) are thought to have been photosynthetic. That sophisticated process arose later and only in specialised groups. If the reverse Krebs cycle always depended on photosynthesis, then it could say little about the origins of life or early evolution – and so little about the deep reasons for the structure of metabolism. The discovery of *non*-photosynthetic bacteria in deep-sea hydrothermal vents that fix CO_2 by way of the reverse Krebs cycle overhauls that conclusion. These bacteria don't need the sun to make their ATP and ferredoxin – they can do it by ancient chemistry alone. We'll see how they pull off that trick in Chapter 4. The point here is that the reverse Krebs cycle pre-dates photosynthesis, and so it might indeed say something important about the deep logic of metabolism.

The second critical discovery relates to the purportedly patchy distribution of the reverse Krebs cycle. Look again at the scheme on page 101. The top step that splits citrate into oxaloacetate and acetyl CoA is not catalysed by the familiar 'citrate synthase' enzyme from our own Krebs cycle running backwards, but by a distinct enzyme called ATP citrate lyase. As it says on the tin, this enzyme uses ATP to split citrate into acetyl CoA and oxaloacetate. The gene encoding ATP citrate lyase was thought to be present in the genomes of all bacteria using the reverse Krebs cycle. In phylogenetic analyses, this gene was taken to be diagnostic for the presence of the reverse cycle; that's the basis for inferring its patchy distribution. But now a study shows that the familiar citrate synthase enzyme really can run backwards in some bacteria, especially at high CO_2 levels. That throws the cat among the pigeons. If the diagnostic gene is unreliable, then it could be that the reverse Krebs cycle is far more widespread across anaerobic bacteria and archaea than had been thought. Right now we have no idea just how widespread it might be. If the reverse Krebs cycle were indeed more ubiquitous in the earliest stages of life, when CO_2 levels were much higher, that could push its significance back to the origin of life itself.

Could it be that the origin of life structured metabolism through the reactions of H_2 and CO_2, to form Krebs-cycle intermediates? That this biosynthetic engine was hard-wired into life from the beginning, the driving force that gave rise to genes and proteins? That billions of years later, as the atmosphere filled with oxygen, the Krebs cycle flipped its direction, creating an extraordinary opportunity to strip and burn organic molecules in a brave new high-octane world? That this reversal placed a tension on the heart of metabolism, which was now obliged to create and destroy from the self-same molecules – a case of having your cake and eating it? That this tension is at the root of ageing and diseases such as cancer, where the fragile balance between energy flow and growth is paramount? Read on.

3

FROM GASES TO LIFE

'Isn't the deep ocean supposed to be like a desert?' Jack Corliss's voice crackled from the depths to the acoustic telephone aboard the *Lulu*. Described as a floating junkyard, the Research Vessel *Lulu* was the original mothership for the submersible *Alvin*. On 17 February 1977, Corliss was leading a three-man team sealed tightly within the *Alvin* two kilometres down, tracking towards the warmth of a suspected deep-ocean hydrothermal vent. The previous day, the cameras attached to a two-tonne underwater cage, towed along the seafloor by a steel cable, had taken 3,000 photographs of bare lava, ostensibly barren of life, before the film ran out. But a tantalising series of thirteen pictures, corresponding to a local spike in temperature, seemed to show an improbable field of clams that appeared abruptly through a cloud of misty blue water, and then disappeared just as quickly. The *Alvin* was sent to investigate, with young Jack Corliss at the controls. The sight that greeted them was unprecedented: tottering chimneys furiously disgorging black smoke into the black depths of the ocean. But that was not what haunted Corliss for decades to come. Responding to his loaded question from aboard the mothership, the research student Debra Stakes duly confirmed that the deep ocean was supposed to be a desert. 'Well, there's all these animals down here', said Corliss.

Exploration of the deep ocean has never quite matched the glamour of space, but from a scientific point of view, this laconic report from the depths was an equally giant leap for mankind. The crew aboard the *Alvin* and *Lulu* on that occasion were geologists, geochemists and geophysicists; nobody had imagined there would be a need for a biologist. The samples brought back to the surface were preserved in a little formaldehyde that a student had happened to bring along, plus a lot of Russian vodka. The names of their extraordinary discoveries were whimsically evocative. Sure, the 'Rose Garden', the 'Dandelion Patch' and the 'Garden of Eden' lacked biological rigour, but no one had seen a field of giant tube worms with feathery red plumes before. The *Lulu* became a voyage of discovery on a par with the moon landings. 'We all started jumping up and down. We were dancing off the walls. It was chaos!', recalled one researcher. This was an unimagined finding: a deep-sea food chain that apparently did not depend on sunlight or photosynthesis, but lived on the gas hydrogen sulfide emanating from the vents themselves.

Such vibrant life so far from the sun swiftly inspired a new perspective on the origin of life. Working with the oceanographers John Baross and Sarah Hoffman, Corliss published a classic paper in 1981, entitled 'An Hypothesis Concerning the Relationship between Submarine Hot Springs and the Origin of Life on Earth.'[1] The trio considered hot, reactive gases emanating from vents across catalytic surfaces containing metals. At the outset they were thinking of gases such as hydrogen, methane and ammonia, which make up the swirling clouds of

---

1 A veteran of many expeditions on the *Alvin*, typically lasting a full day at a time, John Baross touched a chord with me in this disarmingly human observation: 'We have to watch what we eat and drink because there is no bathroom on board.'

Jupiter, and were once thought to have dominated the early atmosphere of the earth too. Back in 1953, in one of the few experiments that ever made it onto the cover of *Time* magazine, Stanley Miller had simulated lightning strikes by discharging electricity through glass flasks containing these gases, to form amino acids, the building blocks of proteins. Corliss, Baross and Hoffman were imagining similar chemistry in their furiously discharging vents. But just as it had become clear that the early atmosphere was probably not steeped in these gases, so too the hotter vents were dominated by relatively oxidised gases, notably carbon dioxide itself, rather than methane and ammonia. It also became clear that the complex food chains surrounding vents were not truly independent from photosynthesis – they were built on the chemical reactivity of hydrogen sulfide with oxygen, the waste product of photosynthesis. The red plumes of the giant tube worms take their colour from a type of haemoglobin, the same pigment that transports oxygen in our red blood cells. The haemoglobin passes its oxygen onto symbiotic sulfur bacteria that live within the body of the tube worm. These bacteria generate their energy and most of the biomass surrounding the vents from the reaction between hydrogen sulfide and oxygen. At the origin of life, before the emergence of photosynthesis, there was little if any free oxygen, so this lifestyle was impossible. And without oxygen there was much less energy in the system.

Yet by the same token, the absence of oxygen should have made the fixation of CO_2 much easier. By the late 1980s the reverse Krebs cycle was finally gaining credence, as gene sequences made it clear that not all autotrophic bacteria depend on the Calvin–Benson cycle. The time was ripe for a new conception of the origin of life. In broad brushstroke, the reverse Krebs cycle uses iron–sulfur proteins, notably ferredoxin, to catalyse the reaction between carbon dioxide and hydrogen, forming carboxylic acids, the carbon skeletons from which all other cellular building blocks are made. The only problem was that the reverse Krebs cycle requires an input of energy (ATP) to work, which in modern bacteria is normally obtained from photosynthesis. Deep down in the bible-black vents, where did the energy come from? Not from photosynthesis, or its waste product oxygen.[2]

Two trailblazing scientists, the chemist Günter Wächtershäuser and the geochemist Mike Russell, independently put forward detailed and contrasting ideas on how iron sulfide minerals in vents might catalyse the hydrogenation of carbon dioxide to form carboxylic acids. Wächtershäuser conceived an elaborate scheme that he termed 'pyrites pulling', which linked the synthesis of iron pyrites (fool's gold) to the reverse Krebs cycle. While ingenious, it bore only a cursory resemblance to the pathway in bacteria. Russell, in contrast, thought in terms of the topology of cell membranes, and specifically the use of proton gradients across such membranes to power growth. I have to say that Mike Russell's ideas came as a revelation to me, and have guided my own thinking over nearly two decades now. He placed his hypothesis in an alternative type of submarine vent, riddled with labyrinths of cell-like pores bounded by thin walls containing iron sulfide minerals. While unknown at the time, such vents were discovered nearly a decade later by Deborah Kelley, captaining the *Alvin*, almost exactly as Russell

---

2 In fact, we now know that deep-sea vents produce enough near-infrared light to power photosynthesis in a few bacteria living in these systems. Some argue that photosynthesis may have emerged in the deep oceans rather than in surface waters scoured by harsh ultraviolet radiation. In any case, photosynthesis is relatively complex and found only in certain groups of bacteria; it's unlikely to have been the primordial energy source.

had predicted. But sometimes details obscure the big picture. Despite some personal feuds (which Freud might dismiss as the narcissism of minor differences) both Russell and Wächtershäuser had envisaged life as originating through autotrophic mechanisms, growing from gases such as CO_2 and H_2 by way of the reverse Krebs cycle. This radical conception opened up a rift of geological dimensions that persists to this day – did life start from CO_2 and H_2 in deep-ocean hydrothermal vents, or in surface waters where ultraviolet light energised gases such as cyanide, as argued by the venerable tradition of prebiotic chemistry going back to Miller?

These ideas are opposed in virtually every respect: deep oceans versus warm ponds, light energy or chemical disequilibria, metabolism first or genes first, autotrophy or heterotrophy, fast chemistry or slow accumulation, local or planetary scales, and biology or chemistry as the ultimate guide to life's origin. Don't worry about the meaning of all these terms for now. Suffice to say that science is often furiously passionate and opinionated until the evidence forces a reluctant agreement, which has not happened yet. I should confess to a personal bias here: as a biochemist with a particular interest in biological energy flow, my own route into the thrilling question of the origin of life links to the first stirrings of biochemistry, rather than the long intellectual tradition of synthetic chemistry.[3] From my point of

---

3 The tradition of synthetic chemistry aims to make the products of interest, such as nucleotides, at high yield, ideally uncontaminated by unwanted products. Synthetic chemists tend to have little interest in the pathways that biochemistry uses today, unless the chemistry seems logical to them (and it often doesn't). Biologists tend to think backwards. Natural selection improves selectivity and yield, so the first pathways had to have lower selectivity and yield, or there would be nothing for selection to work on. But any research programme that specifically aimed for low selectivity and low yield would bring most synthetic chemists out in hives.

view, the discovery of submarine vents was the first time that a geological environment on the early Earth linked meaningfully with the metabolism of known bacteria – a theme that has been developed compellingly over several decades by Bill Martin, one of the most brilliant thinkers on life's origins, who has drawn on metabolism, physiology, genetics and geology to frame the beginnings of biochemistry. My favoured scenario emphasises steady growth – a local environment with unceasing flow, continuously converting CO_2 into organic molecules, which self-assemble into protocells for simple chemical reasons (with membranes forming in a similar way to soap bubbles). It is at least possible to imagine a seamless continuity between geochemistry and biochemistry, life as a product of restless planetary processes. But of course 'it's possible to imagine' is a far cry from 'shown to have happened'. Many a beautiful idea has been killed by ugly facts. How do we go about knowing?

Whatever the answer, these disparate views are far from sterile, but inform practical decisions about where to search for life in the solar system, or further afield. Should NASA and other space agencies back missions to Mars, or to the icy moons of Saturn and Jupiter, Enceladus and Europa? If light is essential for the origin of life, then Enceladus is the last place to look, as those who favour warm ponds are quick to assert. But if life emerges from deep-sea hydrothermal vents, then Enceladus is an ideal place to look, as beneath its icy crust is a liquid ocean bubbling with hydrogen gas and small organic molecules, to judge from the plumes that jet hundreds of miles into space through cracks in the ice. It's the first place I'd look.

Arguably even more important are the practical connotations for metabolism and our own health today. Is the Krebs cycle at the heart of metabolism because life was forced into existence that way, by thermodynamics – fate! – or was this chemistry invented later by genes, just a trivial outcome of information systems that could be rewired, if we are smart enough? Is the difference between ageing and disease an intractable outcome of metabolism, written into cells from the very origin of life, or a question for gene editing and synthetic biology to overcome? That in turn boils down to genes first or metabolism first? The thrust of this book is that energy is primal – energy flow shapes genetic information. I will argue that the structure of metabolism was set in stone (perhaps literally in deep-sea rocky vents) from the beginning. In this chapter we'll explore the evidence for this view, stemming from the origin of life itself.

How old is old?

There is little doubt that the reverse Krebs cycle is ancient, but merely being ancient is far removed from extending right back to the very origin of life. What evidence is there? Simply arguing that the Krebs cycle spins at the heart of biochemistry is not enough. There are two reasons why that might be – either the cycle is genuinely primordial, or alternatively it might be the most efficient network topology, and has therefore been honed by natural selection on genes and cells. These are already complex entities, and by definition arose later than the origin of life. If the Krebs cycle is genuinely primordial it must reflect a favoured thermodynamic path. If it is a product of evolution, refined by selection on genes, then that says nothing about primordial chemistry. Of course, it might be both – thermodynamically favoured, while also being an ideal network topology for metabolism. How can we tell the difference?

The first question is geological plausibility. The reverse Krebs cycle requires the gases CO_2 and H_2 as well as catalysts containing iron–sulfur clusters. Here, we are on relatively safe ground. There was almost certainly a great deal more CO_2 in the atmosphere and oceans around four billion years ago than there is today – probably thousands of times more. Today, most carbon is present as living or dead organic matter, or as carbonate rocks such as limestone. Before the origin of life, plainly there was much less organic matter. You might think that before life there could be no organic matter at all, but the term 'organic molecule' doesn't refer to the molecules made by life, but to the *type* of molecules that make up life – carbon with hydrogen attached. And these can be made by chemists, volcanoes and planets, indeed even asteroids in deep space.

There was also much less limestone on the early Earth, for this tends to form in relatively alkaline oceans. We know from the geological record that barely any limestone was laid down beneath the acidic oceans of the early Earth. Now, if all the carbon in organic matter plus limestone were vaporised and returned to the air and oceans as CO_2, its primordial state, then that would give an atmospheric pressure of about 100 bars of CO_2 (a bar being roughly the modern atmospheric pressure at sea level). Much of this CO_2 would swiftly react with volcanic rocks such as basalt, and so be absorbed into the Earth's crust, but geologists have a hard time accounting for the last 10 bars or so. If only 1 bar lingered, that's still 25,000 times more than today: there was surely no shortage of CO_2. Even when drawn down into the mantle, CO_2 does not remain below for long, but is degassed back into the atmosphere through volcanic eruptions, over a cycle of a few tens of millions of years. The Earth is far from unusual – CO_2 makes up around 95% of the atmospheres of Mars and Venus and is common on exoplanets orbiting other stars.

Hydrogen, too, bubbles from the oceans through the hydrothermal vents advocated by Mike Russell and Bill Martin, which were probably far more common back then. These vents are produced not by volcanic activity, but chemistry – ocean water reacting with minerals such as olivine. Olivine comprises about half of the upper mantle, so when the oceans and the mantle are in direct contact, this rock–water reaction is practically unstoppable. The reaction metamorphoses olivine into serpentine, bestowing the name serpentinisation, which generates strongly alkaline hydrothermal fluids bubbling with hydrogen gas. Today the mantle is insulated from the oceans by the crust, which is mostly composed of relatively frothy, silicate-rich rocks. But four billion years ago the mantle and crust had not yet differentiated, meaning there were no large continents (Earth is a poor name for this water world) and the whole seafloor should have been serpentinising. There's plenty of evidence for extensive serpentinisation in lavas (known as komatiite lavas) dating back to that period. These same reactions are thought to be taking place on Enceladus, giving rise to those plumes of alkaline waters suffused with hydrogen gas and organics. While hydrogen might not have accumulated much in the early atmosphere of the Earth, as it escapes readily to space, there would have been plentiful H_2 in the vents themselves. So the two substrates for the reverse Krebs cycle, H_2 and CO_2, were present at the origin of life in nearly unlimited quantities. That certainly can't be said for the alternative theories based on gases such as cyanide. While relatively stable forms such as ferricyanide might have accumulated in warm ponds on land, there is little evidence for that assertion, and it would not begin to compare quantitatively with H_2 and CO_2.

Of course, having plentiful H_2 and CO_2 is irrelevant if they don't react with each other. This has long been the criticism of prebiotic chemists, who have successfully synthesised amino acids, nucleotides and fatty acids from cyanide and related molecules. In contrast, until recently, the exponents of H_2 and CO_2 had little to crow about; for the most part, these gases remain stubbornly unreactive under relevant conditions. But this conclusion is premature, explained in part through a lack of real effort to search the possible reaction space, including mineral catalysts and higher pressures. In the last few years, there has been a small revolution in prebiotic chemistry – CO_2 has been successfully converted to carboxylic acids, including virtually all Krebs-cycle intermediates. This pioneering work from Joseph Moran and colleagues admittedly used raw iron rather than hydrogen as the source of electrons; but as I write, Moran, working with Martina Preiner and Bill Martin, has successfully used hydrothermal minerals to catalyse the difficult reaction between H_2 and CO_2, forming acetate and pyruvate, critical constituents of the reverse Krebs cycle. Pleasingly, these mineral catalysts include greigite, an iron sulfide with a similar basic structure to the iron–sulfur clusters in ferredoxin, the protein that still catalyses the two most difficult steps in cells today. To my mind at least, it's not coincidental that the Earth provides CO_2 and H_2 in buckets, along with the iron–sulfur catalysts that facilitate their reaction to form carboxylic acids, the same reverse Krebs cycle intermediates that are still at the heart of metabolism today. Thermodynamics and geology are clear: CO_2 and H_2 are ubiquitous, and can react to form Krebs-cycle intermediates.

But other factors are more equivocal. Phylogenetics clearly indicates that the earliest cells in the tree of life were auto-trophic, growing from gases such as H_2 and CO_2 (rather than 'eating' food) but do not really support the contention that the reverse Krebs cycle is ancestral to life, as would be the case if it arose at the origin of life. Besides the reverse Krebs cycle and the Calvin–Benson cycle, discussed in the previous chapter, we now know of another four autotrophic pathways of CO_2 fixation across the living world. Just one of these, the acetyl CoA pathway, is found in both bacteria and archaea, the two great prokaryotic domains of life, implying that this was the only pathway present in their common ancestor, the last universal common ancestor of all life (LUCA). Like the Krebs cycle, the acetyl CoA pathway depends on H_2 and CO_2, but in other respects it seems to be even more ancient – it is a short, linear pathway that does not require any additional input of energy (such as ATP) to fix CO_2. Again, like the reverse Krebs cycle, the acetyl CoA pathway draws on the power of ancient iron–sulfur proteins, including the ubiquitous ferredoxin. All these factors indicate that the acetyl CoA pathway was the ancestral pathway of CO_2 fixation. So, from this phylogenetic point of view, the reverse Krebs cycle seems to be ancient but not ancestral.

On the other hand, the end product of the acetyl CoA pathway is acetyl CoA itself – a C2 molecule that is also part of the reverse Krebs cycle. While acetyl CoA is a lynchpin of metabolism, it isn't the direct source for most other molecules containing more than two carbon atoms. Most amino acids, for example, contain three to six carbon atoms. These are not made (as they could be in theory) by joining C2 units together and chopping bits off as necessary, but from Krebs-cycle intermediates by way of more CO_2 fixation, using more H_2. Likewise sugars. These are formed from the C3 carboxylic acid phosphoenol-pyruvate. And the nucleotides that comprise the hereditary molecules RNA and DNA are made primarily from amino acids and sugars. These pathways are conserved across all life, so are presumably ancestral. In other words, while the acetyl CoA pathway looks to be ancestral, it only solves part of the problem. The remainder of the problem (all the rest of biosynthesis) draws on Krebs-cycle intermediates, even if the reverse Krebs cycle itself does not appear to be ancestral. How do we square this circle?

There are two possible resolutions to the conundrum. The first is that the reverse Krebs cycle is not ancestral, but a few of the constituent carboxylic acid intermediates are. In particular, five of them are universally conserved across life – acetate (C_2), pyruvate (C_3), oxaloacetate (C_4), succinate (C_4) and α-ketoglutarate (C_5). How are these linked if not by the Krebs cycle, you might ask? The answer is quite simple: five out of the six pathways of CO_2 fixation contain Krebs-cycle intermediates. Only the acetyl CoA pathway is conserved in both bacteria and archaea, but it could be that at the origin of life there were multiple ways of making Krebs-cycle intermediates, which eventually 'crystallised' into five different pathways, with distinct network topologies that operate optimally under diverse conditions. These pathways form overlapping groups of Krebs-cycle intermediates.

I mentioned the second possibility in the previous chapter – it could be that the reverse Krebs cycle is conserved more widely than had been thought, in archaea as well as bacteria. That might be true because the enzyme normally used to identify the presence of the reverse Krebs cycle in a genome – ATP citrate lyase – turns out not to be diagnostic for it after all. It seems that the workaday enzyme found in the 'normal' oxidative cycle (citrate synthase, discovered by Krebs himself) operates just as effectively in reverse, at least in some bacteria – and who knows, perhaps most prokaryotes, few of which have been studied under appropriate conditions such as high CO_2 levels. If this reverse operation occurs widely, then the reverse Krebs cycle would indeed look to be ancestral. For now, we just don't know. Do we then give up and wait? Thankfully not. There are other ways to approach the question.

The cycling law

'Energy flows, matter cycles.' This axiom from the legendary biophysicist Harold Morowitz might one day be elevated to a fourth law of thermodynamics, but in the meantime it is more informally known as Morowitz's cycling law. Born in Poughkeepsie, New York, Morowitz was a child prodigy who went up to Yale in the 1940s at the age of sixteen to study physics. Years later, he explained to his own students why he switched from physics to biology. In his freshman physics lab, Morowitz had been paired with another child prodigy, the even-more legendary Murray Gell-Mann, then aged fifteen. The competitive young Morowitz decided that, if Gell-Mann was a typical physicist, then he should change fields. He later declared that outscoring Gell-Mann by a couple of marks in one test was his greatest achievement in physics. I suspect that quote should be interpreted in light of his personal philosophy, conveyed to students in his commencement addresses at Yale (his academic home for decades): 'Conformity is not necessarily a virtue. Hard work is almost never vice. Hopefulness is a moral imperative. And, a sense of humour helps.'

Biologists can be grateful that Morowitz made the switch: he was one of the most profound and delightful of thinkers, combining a physicist's rigour with a biologist's eclecticism. He came to the attention of a wider public in the early 1980s when called up as an expert witness at the McLean versus Arkansas trial, sometimes known as Scopes II, after the Scopes 'monkey trial' in the 1920s. Both trials concerned the teaching of evolutionary biology or 'creation science' in schools. Being a pioneer of non-equilibrium thermodynamics, and with a lifelong fascination with the origin of life, Morowitz testified against the pervasive idea that life breaks the second law of thermodynamics – the tendency for entropy (disorder) to increase in closed systems. The Earth is of course an open system, continually drenched by sunlight, and Morowitz had formally shown that 'the energy that flows through a system acts to organize that system'. The judge, William Overton, ruled that so-called creation science did not qualify as science, so should not be taught in science classes. In his summing up, Overton gave one of the best descriptions of science I've ever read: 'Science is guided by natural law. It has to be explanatory by reference to natural law. It is testable against the empirical world. Its conclusions are tentative (not necessarily the final word). It is falsifiable.' None of these criteria apply to creation science or its prodigal offspring, intelligent design, which call on miracles and the supernatural to 'explain' the natural world.[4]

Morowitz had laid out his cycling theorem in a 1968 book, *Energy Flow in Biology*: 'In steady state systems, the flow of energy through the system from a source to a sink will lead to at least one cycle in the system.' The idea made a pleasing link between living systems (where an organism remains in a steady state, even though all its parts continuously change) with familiar natural phenomena such as tornadoes, hurricanes,

---

4 Being by then something of a scientific celebrity (and by no means unsympathetic to religion), Morowitz wrote a monthly essay for *Hospital Practice* for twenty-two years. These eclectic essays range widely across science and society, and were published as five collections with playful titles such as *The Thermodynamics of Pizza* and *The Kindly Dr Guillotine*. C. P. Snow hailed them as 'Some of the wisest, wittiest and best informed that I have read.' Morowitz acted as a consultant to NASA on various missions, including the Viking mission that probed the surface of Mars for signs of life, was an advisor on the Biosphere 2 project, the largest closed ecological system ever created (designed to explore the viability of closed systems to support human life in outer space), and was one of the founders of the famous Santa Fe Institute. A life well spent. He even managed to complete a magnificent tome on the origin of life, with Eric Smith, which was published months after he died from sepsis at the age of eighty-eight in 2016.

anticyclones and whirlpools, and more obliquely, galaxies and ecosystems. Morowitz saw nutrient cycles in ecosystems as a necessary outcome of energy flow. For example, he noted that photosynthesis makes organic molecules (and the waste product oxygen) from water and CO_2, whereas respiration uses oxygen to break down those same organics, returning them to CO_2 and water. Overall these processes cycle between the two different states, following distinct mechanistic pathways from CO_2 to organic molecules and back. A requirement for distinct mechanistic pathways applies not only to photosynthesis and respiration, but also to the physical chemistry underpinning them, which can happen under sterile conditions too. A simple reversal is not possible, for much the same reasons that, on a planetary scale, the Earth absorbs high-energy sunlight while emitting low-energy heat, remaining in overall energy balance – another great steady-state cycle. The bottom line is that the dissipation of heat means that no pathway is perfectly reversible, hence energy flow powers the cycling of matter.

Always an admirer of the iconic metabolic charts of biochemistry, Morowitz could not help but see the Krebs cycle as the perfect cyclone of matter. There it is, at the centre of all biochemical charts, linked fundamentally to energy flow. Although the reverse cycle had been discovered several years before his 1968 book, Morowitz had not at that point integrated it into his thinking; but later on he came to see the reverse Krebs cycle as being dictated by thermodynamics. If energy flow through a system acts to organise that system through the cycling of matter, and if this law applies as much to metabolism (and its origins) as everything else, then the Krebs cycle seems inevitable, ordained by the irresistible god of thermodynamics.

Even better than that. As noted in the last chapter, the reverse Krebs cycle is autocatalytic. Starting with one molecule of the C4 oxaloacetate, one spin of the reverse cycle generates two molecules of oxaloacetate; the next spin generates four molecules, then eight, sixteen and so on. Not only is this exponential growth, but each doubling forms a copy of the original, combining stability with growth.[5] No wonder Morowitz was entranced. And it just kept on getting better. The idea of an autocatalytic cycle that turns CO_2 into organic molecules, dictated by the laws of thermodynamics, was lent even more force by the repetition of simple, almost mundane, chemistry in the reverse Krebs cycle. This repetition does not occur in our shorter oxidative pathway (or in any of the other five pathways of CO_2 fixation) but is manifest in the reverse Krebs cycle. Mundane repetition hints at inevitability – it is so deeply etched into the fabric of nature that it just can't stop happening, over and over again (see figure on the next page).

Inevitable versus implausible

Morowitz's conception is beautiful and full of meaning. It suggests that life can't help but emerge on any wet, rocky planet flooded with light from a nearby star. He even wrote a book, *The Emergence of Everything*, which explores the limits of inevitability – where does the determinism of chemistry give

---

5 Just in case you are confused ... the normal 'forwards' Krebs cycle is catalytic, which is to say one complete spin of the cycle will regenerate the starting point. If you add succinate, you will regenerate exactly the same amount of succinate after a turn of the cycle, while making more of the end products CO_2 and 2H. In contrast, the reverse Krebs cycle is autocatalytic. This means that a complete spin of the cycle regenerates the starting point in two copies. If you start with succinate, one complete spin of the cycle will generate two molecules of succinate, and so on. That allows an exponential lift-off, or alternatively, it regenerates the starting point, plus another Krebs-cycle intermediate that can be used for other purposes such as amino acid synthesis.

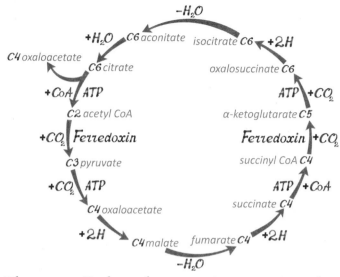

*The reverse Krebs cycle repeats its succession of steps,
highlighted in bold lettering. Starting with ferredoxin the
repeated steps are: $+CO_2$, $+CO_2$, $+2H$, $-H_2O$, $+2H$, $+CoA$.
The only difference on the second rendition is that the final
$+2H$ becomes $+H_2O$. Notice that the requirement for ATP at
the critical steps near ferredoxin is also mirrored. Each full
turn of the Krebs cycle produces one additional molecule
of oxaloacetate, along with the oxaloacetate in the cycle
itself, driving stable growth through an autocatalytic loop in
which one copy begets two, allowing exponential take-off.*

way to the wiles of biology? But beauty can be a treacherous
guide in science, as in life. Life is rarely simple and there are
some grave problems with Morowitz's view too. Perhaps the
most serious was pointed out by another pioneering origin-of-
life researcher, Leslie Orgel.

Orgel had trained as an inorganic chemist at Oxford in
the 1940s. He was among the first to see the molecular model
of DNA constructed by Crick and Watson in 1953, when the

great crystallographer Dorothy Hodgkin told a group of her colleagues that they had to cram into two cars to make the trip to Cambridge. Crick later joined Orgel at the Salk Institute in San Diego during his brief foray into the origin of life in the 1970s, which ended with Crick's rather strange book *Life Itself*, on the possibility of directed panspermia – the idea that life was deliberately seeded on Earth by an alien civilisation. That conception grew from the problem of the genetic code, which Crick had pronounced a 'frozen accident', meaning that the code was essentially random, and became universal not because it was better than any other code, but because mutational changes to any code, once established, would be catastrophic, hence could not occur. We now know that's not correct. Even so, these ideas gave rise to the 'RNA world', pioneered by Orgel among others. This hypothesis argues that RNA can act as both a template and a catalyst, and so could have given rise to both heredity and metabolism at the origin of life.

Never shy of confronting the difficulties implicit in his own ideas, Orgel observed that if metabolism were indeed invented by an RNA world, then the structure of biochemistry could give no insight into the origin of life. That's because metabolism would be a product of genetic information – written in the genes – and not thermodynamic necessity in some favourable geochemical context. In other words, genes came first. Information rules biology, and whatever form of prebiotic chemistry came before genes (said Orgel) could not be read by looking at the products of genes. That would include the reverse Krebs cycle, which Orgel argued had to be a product of genes and selection, and so could give no insight into prebiotic chemistry.[6]

---

6 That of course raises its own questions about just how RNA acquires information, a problem so severe that thinkers of the calibre of Paul Davies and

Famous for pithy remarks such as his eponymous Second Law, 'Evolution is cleverer than you are', Orgel memorably dismissed self-organising biochemical cycles as 'an appeal to magic'. He returned to the theme in a final testament, published posthumously, entitled 'The Implausibility of Metabolic Cycles on the Prebiotic Earth'. In his closing lines, he scorned the prebiotic reverse Krebs cycle as 'if pigs could fly hypothetical chemistry'. He saw two serious, interconnected issues with the reverse Krebs cycle in the absence of genetically encoded enzymes: side reactions and yield. By speeding up specific reactions, enzymes funnel chemistry down particular pathways, limiting side reactions, and therefore dissipation to unwanted products. Take any single step in the reverse Krebs cycle: it's just one of several possible reactions, and not necessarily the most likely of them. In the absence of enzymes, only a small proportion of these scattered products will be the ones needed for the next step in the pathway. This means that the yield falls off, step by step, approaching zero after a handful of steps. Orgel considered it literally incredible that the reverse Krebs cycle, which takes twelve successive steps to regenerate its starting point, could possibly do so without enzymes increasing the selectivity and yield of each step. Then came the sucker punch. The first step of any cycle depends on the yield of the final step. Fat chance of autocatalysis. Each cycle would have precisely nothing to amplify.

There's no question that Orgel had a powerful point. I think that it's fatal to the idea of the reverse Krebs cycle – as a strict cycle – being truly prebiotic. On the other hand, there is clearly

---

Sara Walker have argued we need a new law of physics to explain the origin of information. It seems to me that this is less of a knotty problem if energy flow structures the origin of genes and information; we'll come to how that might work later in the chapter.

something favoured about the Krebs-cycle intermediates. They form spontaneously from CO_2 and H_2, given the mineral catalysts and high pressure, to the exclusion of many other possible products – chemistry that had not yet been demonstrated in the lab before Orgel died. But equally importantly, they are not formed as a complete cycle, so Orgel's point stands. And there is another problem with a complete cycle. The side products are not unwanted at all: they make up the rest of metabolism. We *want* to form amino acids, fatty acids and sugars from Krebs-cycle intermediates – they are the real building blocks of cells. These, too, seem to be thermodynamically favoured products. Markus Ralser and colleagues have shown that other central metabolic pathways, including gluconeogenesis and the pentose phosphate pathway, occur spontaneously in the absence of enzymes. Just to give a feel for how far this favouritism goes, a PhD student in my lab, Stuart Harrison, is currently working on the synthesis of nucleotides from amino acids and sugars, following a prebiotic version of the metabolic pathway. Most of the steps do indeed form the right products, given simple catalysts such as metal ions. The difficulty is getting all the steps to join up, but he is making exciting progress.[7]

My point here is more specific. The last thing that prebiotic

---

7 The picture emerging is reminiscent of the anthropic principle – the idea that the cosmological constants (such as the strength of gravity) had to be very similar to their measured values for the universe we know to exist at all – for stars, planets, organic chemistry and life as we know it to be possible. They are 'finely tuned'. Perhaps there is a multiverse in which different universes have different cosmological constants, and we have to live in one that was capable of giving rise to us; or perhaps there is some deeper reason, buried in the laws of physics, for the cosmological constants to be what they are – a theory of everything. In similar vein, it looks increasingly as if carbon chemistry is bound to give rise to the core metabolism of life as we know it. What that might say about the nature of the universe is another question, and before we go anywhere near it, we need to establish whether it is indeed true, or merely reflects some bias.

chemistry needs is the perfect autocatalytic cycle, which makes copies of itself at the expense of all the side reactions that make up living cells. But a cycle that bleeds off material with every spin ceases to be autocatalytic, and so becomes far less beguiling to those who dream of an exponential lift-off in bacteria or biochemistry. It is just a cycle that probably can't make it all the way around. A broken cycle.

Where does this leave us? Compromise can be a dirty word in science, but it seems to me that there really is a pattern of mundane, repetitive reactions that form Krebs-cycle intermediates in the right order, but initially just the first few steps, perhaps up to the C3 or C4 intermediates. Not a cycle, but a straight line, let's call it the Krebs line. We've seen that there's strong experimental evidence, from the last few years, that these Krebs line intermediates do indeed form spontaneously from CO_2 and H_2. That's a big deal, because it substantiates a hypothesis that already has unique explanatory power in connecting an omnipresent geological environment – alkaline hydrothermal vents – to the heart of biochemistry, the carboxylic acids from which other cell components are made. Experimental work shows that some of those cell components (some amino acids, sugars and fatty acids) really can be made from Krebs line intermediates under equivalent conditions.

In principle, this means that the sustained disequilibrium in hydrothermal vents – the non-stop replenishment of both H_2 and CO_2 – could drive a continuous reactivity, in which the core components of biochemistry are continually formed and react together to form more complex networks of reactions. So continuous flow would drive, if not an exponential lift-off, at least continuous growth. But for this scenario to be workable depends on the rate at which new organics form from H_2 and CO_2 relative to the rate at which they dissipate – get washed away or diluted or break back down. And that rests, first and foremost, on the chemistry itself.

The likelihood of any reaction taking place depends in part on the stability of the products, but most of all on the proclivity of molecules to react at all – the kinetics. In chemistry, the proverb 'Where there's a will there's a way' is mistaken, because sometimes there just isn't a way. Not all chemical paths have a will and a way. At the origin of life, we are seeking a perfect storm, in which the right reactions are driven by the right catalysts in the right place at the right time. If that sounds like a tall order, one factor alone shows that it's not: mineral surfaces. For me, the chemistry of the Krebs line on minerals unites the will and the way so tightly that I want to exclaim, with Matthew, 'Strait is the gate and narrow is the way which leadeth unto life.' This is simple chemistry that revolves around what happens when carbon dioxide binds to mineral surfaces in the presence of hydrogen. If you've been wondering why the Krebs cycle is so central to life, this chemistry is your answer: by continually transforming the simplest, most abundant gases into the core molecules of life, the Krebs cycle (as it eventually becomes) really does bring life alive.

Magic surfaces

Consider what happens close to a charged surface, let's say a mineral containing iron and sulfur (such as greigite), shown below as a wavy line to keep things simple. Imagine that a CO_2 molecule comes into close apposition with the surface and is held in place by electrical attractions between the oxygen atoms and the mineral surface, depicted here with dashed lines. I'm saying 'imagine it', but let me be clear: experiments show that CO_2 adsorption and reduction, as I describe below, does indeed happen. Binding to a surface works a bit like an enzyme, feeding electrons continuously and positioning molecules 'just so' for the next steps to happen. Let's walk through the steps, which I've made as friendly as I can through my 'portraits' of the protagonists. I should say that, by convention, a curly black arrow denotes the movement of a pair of electrons, while the dotted lines represent electrical attractions between oxygen atoms and the surface.

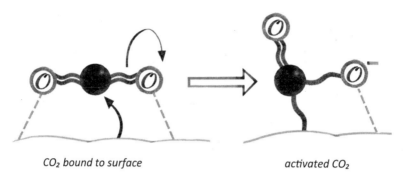

CO_2 bound to surface   *activated CO_2*

What just happened here? Look at the curly arrows again. A pair of electrons passes directly from the surface onto the carbon atom, which promotes the transfer of a bonding pair of electrons onto the oxygen, giving it a negative charge. Because one of those electrons belonged to oxygen in the first place, by claiming the pair it only gains a single negative charge. I appreciate that chemical mechanisms can be intimidating to look at, and frankly they're impossible to draw, forcing a trade-off between clarity and reality. Most of the individual steps I'm going to show here are not 'real', as the whole series occurs virtually instantaneously in a sort of soft-shoe shuffle. For clarity, I'll split up the mechanisms into one or two steps at a time, the penalty being there are then pages of them, making the overall process look more difficult than it really is. But the benefit is that pictures bring the inexorable logic of each step to life in a way that words cannot.

The structure above is stabilised by the transfer of electrons from the mineral surface, in much the same way that ferredoxin stabilises the transfer of electrons onto CO_2 in cells today (see Appendix 1). The parallel between mineral surfaces and enzymes such as ferredoxin is remarkable: the lattice structure of iron and sulfur in minerals such as greigite is nearly identical to the tiny clusters of iron and sulfur that perform the chemistry in ferredoxin. Both the minerals and clusters can also incorporate other metals that transfer pairs of electrons even more readily, especially nickel. So how far does this parallel between enzymes and mineral surfaces go? Might it even explain the repeating chemistry of the Krebs cycle? Let's see what happens next.

Now another electron jiggle splits the bound CO_2 into carbon monoxide (CO) and a charged oxygen atom, both still attached to the surface. Knowing exactly how electrons behave during these jiggles is a quantum artform in itself, uncertain in detail even today – electrons 'tunnel' (hop) across these tiny distances and simply appear where they are most likely to be – but experimentally there's no doubt that CO_2 binding to a surface is reduced to CO, roughly as depicted here:

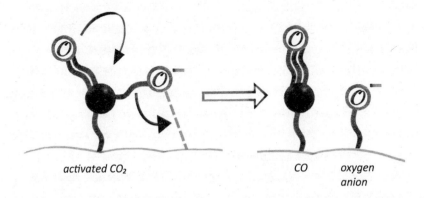

activated CO_2 CO   oxygen anion

All this chemistry is promoted by *electrons* passing from the surface of the mineral onto CO_2. But as in respiration, protons and electrons can unite to form hydrogen atoms. This happens much more readily in mildly acidic environments. Under these conditions, protons can bind to sulfur onto the surface of the mineral, positioning them close to the action. They can then follow the electrons onto CO_2 attached to the surface. Overall, the transfer of electrons and protons onto CO_2 equates to the transfer of hydrogen, even in the absence of hydrogen itself. Here's a sequence showing roughly what happens:

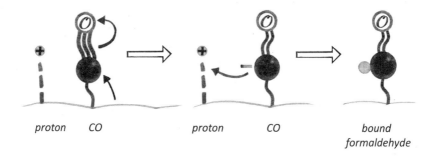

proton   CO   proton   CO   bound
                                            formaldehyde

As before, the curly arrows depict the movement of a pair of electrons, initially from the surface of the mineral onto the carbon atom, allowing the oxygen atom to take back the electron pair it had previously donated to form the triple bond of carbon monoxide. As you might imagine, a negative charge on the carbon atom, in the middle of the panel, is not a stable arrangement, but in a mildly acidic environment, where protons are likely to bind nearby, it lasts for the merest instant, before a proton hops across, and balances charges. The convention of the curly arrow can be a bit confusing here if you're not used to it. It doesn't mean that the pair of electrons physically leaves the carbon atom and crosses the gap to join onto the proton, as seems to be indicated by the arrow. Rather, it means the electron pair is shared with the proton, while remaining firmly attached to the carbon. If you prefer, you could imagine the electrons crossing to the H^+ to form H^- (a 'hydride' ion) leaving a positively charged

carbon behind. What would happen next? The H⁻ would immediately attack the C⁺, forming a covalent bond. Either way, the
end product is a hydrogen atom attached to a carbon.

There's an even more important general principle here,
which we will return to in a while: it's much easier to pass
electrons onto carbon atoms in an acidic environment, where
protons are abundant, than it is in an alkaline environment,
where there are hardly any around. Because a proton plus an
electron equates to a hydrogen atom, the overall effect is to add
a hydrogen atom onto the carbon, giving an organic molecule.
The same thing can happen repeatedly. So here's what happens
next. Notice in the middle panel that the electron pair on the
oxygen is quickly shared with a nearby proton, in this case
forming an OH (an 'alcohol' group):

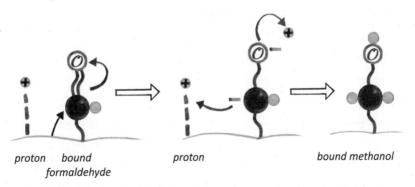

proton   bound   proton   bound methanol
         formaldehyde

The final step in this sequence is another repeat, but this time
the oxygen takes its final bonding pair of electrons and detaches
as a hydroxide ion (OH⁻), leaving now a methyl group (-CH₃)
bound to the surface. This is the same stout belly that we've
seen in pyruvate or acetate. In other words, what started out
as a fully oxidised carbon, bound to two oxygen atoms in the
form of CO_2, ends up bound to multiple hydrogen atoms (it is
said to be 'reduced') still attached to the surface. That final step
goes like this:

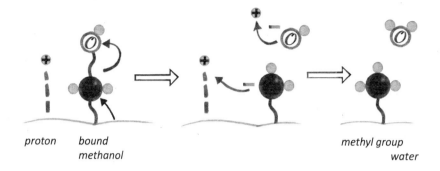

proton   bound   methyl group
          methanol   water

Notice what else is happening here. The OH⁻ has detached in a mildly acidic environment, so there are plenty of H⁺ ions around. These react immediately to form water, H_2O, shown in the right-hand panel – a fizzing neutralisation reaction that many of us might remember from titrating acids and bases at school, perhaps even in your mouth. This is important for two reasons. First, it means that the OH⁻ ions detach easily in acidic environments, but not under alkaline conditions, where there's already an excess of OH⁻ ions in the surroundings. And second, look what's formed overall – water! This is what's called a dehydration reaction: water is being removed in a watery environment. It's sometimes said that this kind of dehydration is impossible at the origin of life, but of course cells do it all the time, albeit using enzymes and ATP. A deep secret of life is that enzymes never pull water from a hat. Rather, the constituent parts of water are removed, OH⁻ and H⁺, which can combine to form water, as in this case, or may simply attach themselves independently to other molecules such as phosphate. The key point here is that the transfer of electrons in an acidic environment actively favours the elimination of water – dehydration in a wet environment.

And now for the most startling step. I have to admit I'm amazed that this happens, but it certainly does, as has been known since the 1920s when Franz Fischer and Hans Tropsch pioneered a method for generating synthetic gasoline from coal. This 'Fischer–Tropsch synthesis' later fuelled the German war effort, Germany being rich in coal but poor in oil. The details of the original industrial process need not concern us here (as it happens at high temperatures and pressures in the gas phase, from CO) but much the same processes have been shown to happen under hydrothermal conditions. Here's what happens in our setting, when a methyl group is positioned next to a CO:

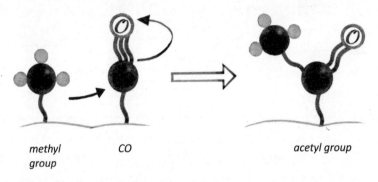

methyl   CO   acetyl group
group

The CO, recall, was our starting point, formed when CO_2 binds to the mineral surface and is split into CO and a negatively charged oxygen atom. The methyl ($-CH_3$) group, which is formed as we've seen through the transfer of electrons onto an earlier CO molecule, can now hop across to bind onto the CO. That's the part that I find surprising, although evidently it happens, both in the industrial Fischer–Tropsch process as well as in cells – the ancient, widespread iron–nickel–sulfur enzyme carbon monoxide dehydrogenase catalyses exactly this reaction. But now look at the product. This bound molecule is a prebiotic equivalent to acetyl CoA – discovered, if you recall, by Fritz Lipmann, and one of the most important molecules in all metabolism. We have just pulled it out of a hat by a simple trick. Rather than being bound through the sulfur to the handle that is coenzyme A, in this case the acetyl is bound directly to the surface. Critically, it shares the same relatively high reactivity of acetyl CoA. And it can be released as free acetate easily enough by reacting with a negatively charged oxygen atom (which is also formed in the first step, as shown on page 135):

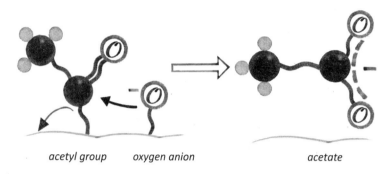

acetyl group   oxygen anion   acetate

Notice that I'm depicting an electron pair returning to the surface here. Iron–sulfide minerals are semiconducting, which is to say they can transfer electrons across themselves. This means that electrons transferred from the surface onto CO_2 can be replenished, so the process can happen over and over again without any permanent change in the surface – the surface is catalytic. Every time a pair of electrons transfers from the surface, the iron or nickel in the surface becomes oxidised, so it loses electrons. If the missing electrons are not replenished from elsewhere, then the electrons can return, as shown above, from carboxylic acids bound to the surface, detaching the bound molecule, in this case as free acetate. That reaction is very close to equilibrium and can go either way.

Conversely, if the electrons are replenished from elsewhere, then this chemistry can go on and on in remarkably similar steps. If you don't believe me, look at Appendix 2, where I've drawn out the first half of the reverse Krebs cycle step by step, following the same ground rules. As Morowitz observed, the second half repeats the same chemistry as the first half. No wonder these molecules are beginning to crop up in experiments that start with CO_2 and minerals. But where exactly are all these electrons being replenished from? The answer is simple: hydrogen gas, bubbling straight out of alkaline hydrothermal vents. If the concentration of hydrogen in hydrothermal fluids is high, that tends to favour the continuous transfer of electrons from the surface, forming not only Krebs-cycle intermediates, but also longer-chain carboxylic acids – fatty acids – capable of forming membranes. On the other hand, if the concentration of hydrogen is lower, then smaller carboxylic acids are more likely to be released from the surface, as electrons flow back onto the barrier. So the ultimate source of electrons to fix CO_2 is H_2. And that comes with its own requirements for cellular structure.

The driving force

Why has it been so difficult, experimentally, to persuade H_2 to react with CO_2, at least until recently? There are three aspects to this problem: catalysts, pressure and pH (proton concentration). We've already discussed catalysts. Iron–sulfide minerals and other surfaces promote the right chemistry, as laid out over the last few pages. To act as a catalyst, though, requires a continuous replenishment of electrons from H_2. The problem here is that H_2 is not very reactive. It has lukewarm chemical desire to push its electrons onto any molecule that doesn't have a fierce craving for them (such as oxygen). This problem is exacerbated in solution because H_2 is poorly soluble, so in water there isn't a lot of it around. The likelihood of a reaction taking place depends in part on the availability of the reactants. If that is low, there's less chance of the reaction happening. That's why pressure is so important: the more intense the pressure, the better hydrogen dissolves, so in effect the more reactive it becomes. The successful experiments that persuaded H_2 to react with CO_2, using iron–sulfide catalysts, worked best at pressures of around 100 bar, equivalent to an ocean depth of about 1 kilometre. If life did start in deep-sea hydrothermal vents, then pressures of hundreds of bars would be typical, which of course might in itself favour deep-sea hydrothermal environments as the incubators of life.

But living cells can thrive on H_2 in surface environments at low pressure. Life doesn't work miracles, so how do cells pull off that trick? The explicit problem for cells is: how do they coax trace amounts of H_2 to transfer their electrons onto ferredoxin, which then passes them onto CO_2? There are several ways of doing that, but most of them are complex, and involve multiple enzymes, so don't seem relevant to prebiotic chemistry. The most intriguing possibility, which connects cells to alkaline hydrothermal vents, is that the reactivity of both H_2 and CO_2 depends on the local concentration of H^+ ions, which is to say, pH. We've noted this already for CO_2. Many of the steps needed to convert CO_2 into carboxylic acids work better at mildly acidic pH, as protons balance the charge of electrons and promote dehydration reactions in water. Paradoxically, just the opposite holds true for H_2 gas. H_2 is more reactive in alkaline conditions, where protons are scarce, while OH^- (hydroxide) ions are abundant. The reason is simple enough. Imagine that H_2 passes its electrons onto a catalyst, be that ferredoxin or an iron-sulfide mineral. H_2 is composed of two electrons and two protons, so when H_2 transfers its electrons onto the catalyst, a pair of protons is left behind. Clearly, that is not favoured in an acidic environment, already chockful of protons, which would become even more acidic. This sort of thing is opposed by thermodynamics. But if the environment is alkaline, then the H^+ released can react immediately with OH^- to form water, in a fizzing neutralisation reaction. You'll never stop that happening. The long and short of it is that H_2 will push its electrons onto other molecules (such as iron–sulfur catalysts) far more aggressively in alkaline conditions.

So the dilemma here is that H_2 will push its electrons onto the catalyst in alkaline conditions, but CO_2 will only accept them from the catalyst in acidic conditions. That's why the reaction doesn't happen easily at any uniform pH, and instead requires high pressure to ramp up the reactivity of H_2 gas. *But the chemistry of cells does not take place at uniform pH.* Far from it. Virtually all cells pump H^+ out, making the outside about three pH units more acidic than the inside – which is to say, there's a thousand-fold difference in H^+ concentration.

In ancient cells such as methanogens, the influx of H^+ through iron–sulfur proteins associated with the membrane promotes the transfer of electrons from H_2 to CO_2 to form organics. The deepest requirement for the proton-motive force might therefore be CO_2 fixation. The prime example is the 'energy-converting hydrogenase' or Ech. This membrane protein has four iron–nickel–sulfur clusters, which transfer electrons from H_2 to ferredoxin. Two of the clusters sit right next to a proton channel in the membrane, and their properties depend on proton binding, which is to say, the local pH. So, when Ech binds protons, it can accept electrons from H_2 (in the jargon, it is more easily reduced). And when the protons detach, Ech becomes more reactive, and can now force its electrons onto ferredoxin, which in turn pushes them onto CO_2. Then incoming protons bind Ech again, and the cycle repeats itself. In other words, Ech acts as a switch that extracts electrons from H_2 when in 'oxidising' mode (with protons bound) and forces them onto ferredoxin when in 'reducing' mode (deprotonated). The difference between the two modes is about 200 millivolts, so the simple pH gradient makes the 'impossible' not just possible but virtually unstoppable. And unlike molecular machines such as the ATP synthase, this mechanism could work in prebiotic conditions.

Here's how. Alkaline hydrothermal vents are riddled with interconnected pores that have a similar topology to cells – alkaline inside, acid outside. H_2 gas is delivered in alkaline hydrothermal fluids, making it more reactive, whereas CO_2 comes dissolved in acidic ocean water, making it more willing to accept electrons. In early vents (lacking dissolved oxygen) the two phases were separated by thin barriers containing iron sulfide minerals, which could transfer electrons from H_2 on one side to CO_2 on the other. The theory is soundly grounded, but that's not enough. This is an experimental question, which my lab and others have been struggling to address for years now. We've had limited success, because H_2 is difficult to work with under continuous flow at high pressure. But as I write, Reuben Hudson and Victor Sojo (a former student of mine, who generously included me among his collaborators) have shown that pH gradients across thin inorganic barriers do indeed facilitate the transfer of electrons from H_2 on one side, to CO_2 on the other, at a pressure of just 1.5 bar. This was proved beautifully using isotopes of carbon and hydrogen to show that the electrons did indeed come from H_2, whereas the protons derived from the acidic ocean solution, as laid out here. And crucially, the carbon in the organic molecules (mainly formate) really did derive from the CO_2 added. No organics were formed at normal air pressure, or without H_2 gas, or in the absence of a steep pH gradient. This is beautiful experimental verification – for once, not slain by an ugly fact. I take my hat off to them.

Small steps, giant leaps

How do we get from pores in hydrothermal vents to the first cell-like structures? You can be forgiven for thinking that these first few steps of prebiotic chemistry are still a long way from life, but in some respects they are closer than you might think. For example, fatty acids, the main constituents of cell membranes, are simply longer-chain carboxylic acids, which can be formed through much the same sort of chemistry, as noted earlier. These, too, have been made by Fischer–Tropsch synthesis under hydrothermal conditions. Likewise, some amino acids can be formed easily from carboxylic acids by substituting a nitrogen for an oxygen atom, or vice versa (as Krebs had found back at the beginning).

Together, these products can take on amazingly life-like properties. Simple mixtures of fatty acids form spontaneously into what can be called 'protocells', with a thin bilayer membrane similar to modern cell membranes enclosing a watery internal volume. These are mesmerising entities that fuse together and divide in two, with every appearance of bustling purpose, driven only by heat in the environment. Think about the magic of dancing soap bubbles and you'll get the general idea. Working in my lab, Sean Jordan and Hanadi Rammu have shown that these protocells self-assemble even more readily under hydrothermal conditions, such as pH 11, 70°C and ocean salinity. In just a few small steps of quite repetitive chemistry, we've leapt from strictly inorganic pores to organic protocells, bounded by lipid membranes, poised to … do what exactly?

The next step is just as pleasing. How could these protocells grow and divide? To do so, they would need to make more organic molecules inside themselves (more fatty acids, more amino acids) using the proton gradient. The million-dollar question is: could a prebiotic version of Ech operate in protocells, driving their growth? The short answer is yes. We were amazed to find that when mixed with iron and sulfide in solution, the amino acid cysteine spontaneously forms exactly the same type of clusters found in both Ech and ferredoxin (technically called 4Fe4S clusters). They even have a similar tendency to force electrons onto other molecules such as CO_2. These iron–sulfur clusters could in principle drive the formation of new carboxylic acids, fatty acids and amino acids *inside* the protocells – which is to say, they could drive protocell growth.

If you are unusually alert, you may have noticed a discrepancy between organic synthesis in inorganic pores (where I talked about *electrons* crossing inorganic barriers) and organic synthesis in protocells (where I talked about *protons* crossing cell membranes). The fundamental requirement is that H_2 is in alkaline conditions, while CO_2 is in acidic conditions. In principle, this arrangement could be achieved by protons crossing the inorganic barrier, rather than electrons. In fact, that is feasible. We have shown that protons cross iron-sulfide barriers two million times faster than hydroxide ions going the other way. That discrepancy can generate steep pH gradients just within the alkaline side of the barrier. In this case, 'steep' means four pH units (a 10,000-fold difference in concentration) across iron-sulfide nanocrystals barely 25 millionths of a millimetre (25 nanometres) in diameter. If so, then inorganic pores and protocells are potentially homologous. Because protocells stick to iron-sulfide barriers on the alkaline side, protons could flow straight across the barriers and into the protocells lining the barrier, before being dissipated by the alkaline hydrothermal flow, giving a beautiful continuity between geochemistry and biochemistry.

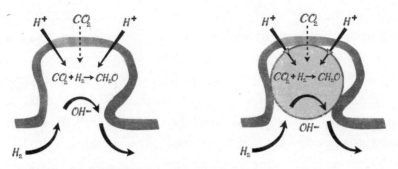

Alkaline hydrothermal fluids

Transfer of protons across an inorganic barrier (left) or the same barrier enclosing a protocell (right) could drive the formation of organics (CH_2O) within the pore or the protocell. The triangles in the membrane of the protocell (on the arrows) symbolise iron–sulfur clusters bound to amino acids, associated with the membrane – a proto-Ech (see text). Growing protocells must remain bound to the barriers in hydrothermal vents to tap the geologically sustained proton-motive force. CO_2 crosses the barriers much more slowly than protons, hence the dotted arrows.

Perhaps the most important property of this system is that it can in principle make copies of itself, and over time can become more complex. For reasons of physical chemistry, new fatty acids will go straight into the membrane, whereas a proportion of amino acids should interact with iron and sulfide, generating more ferredoxin-like FeS clusters. The protocells with the most FeS clusters should make the most new organics, favouring more of the same physical interactions. That would generate more FeS clusters, so more organics, and so on. In other words, protocells with more FeS clusters will grow faster, and pass on more FeS clusters to their daughter cells. This kind of positive feedback loop is a very direct form of physical heredity. It's hardly an edifying thought but the first principle of heredity is: to him that hath shall be given. In any case, the continuous flow of H_2 and CO_2 in a structured environment should drive the replication of protocells, and heredity begins with an unequal form of growth.

Positive feedbacks can take us further. We've already seen that much of core metabolism just springs into existence in the right conditions. The first 'biological' catalysts must have sped up 'helpful' aspects of geochemical flux. But helpful to what? Helpful to whatever is being replicated – protocells in this case. So the first biological catalysts sped up protocell growth, meaning ultimately the conversion of H_2 and CO_2 into the fabric of new protocells. For my money, the first nucleotides, and eventually RNA and DNA, emerged inside such replicating protocells through positive feedbacks. Nucleotides still function alongside enzymes today, and can catalyse important reactions, notably CO_2 fixation, and the transfer of 2H in general. Just think of NADH, which we've seen is responsible for transferring 2H in plenty of biological reactions. Recall that the acronym stands for *nicotinamide adenine dinucleotide* ... it is a nucleotide. So, as soon as prebiotic pathways for nucleotide synthesis arose (facilitated by higher concentrations of precursors and catalysts in growing protocells), they will contribute to protocell growth through CO_2 fixation and hydrogenation – a positive feedback.

Raquel Nunes Palmeira, Stuart Harrison and Aaron Halpern, three PhD students in my lab, are modelling how the genetic code might have emerged in this setting.[8] Assuming that replicating protocells do indeed make nucleotides, the next step is to polymerise them into random sequences of RNA. These have zero information content. You might think that means

---

8 That wasn't the original plan, but cracking the code is exciting and we're on a roll. How could I stop them?

zero utility too, but that's not so. Think about random genetic sequences in terms of protocell growth. Only strings of RNA that enhance protocell growth would be favoured, while RNA that impedes protocell growth (through 'selfish' behaviour) would be selected against.[9] Meaning emerges with function. In other words, the genetic code had no need to 'invent' information from nothing: information took its meaning from protocell growth from the beginning. From this point of view, genetic information simply enables a more exact form of growth: genes reproduce their own system more accurately and rapidly, which is to say they help protocells get better at copying themselves. So long as replicating protocells can arise spontaneously – and they can – then there is no conceptual problem with the origin of information.

All this implies that genes arose in protocells, and their initial value was promoting growth from H_2 and CO_2 by way of Krebs-cycle intermediates. This requirement for growth at

---

9 You might be wondering how a random string of RNA could affect protocell growth. The answer depends on biophysical interactions between RNA and amino acids. There is an intriguing 'code within the codons', which suggests, for example, that hydrophobic amino acids will preferentially interact with hydrophobic bases in the RNA triplet codon. We can predict quite a lot of the genetic code through biophysical interactions involving only the size and hydrophobicity of amino acids. So a short sequence of RNA containing mostly hydrophobic letters (such as G and A) would interact with hydrophobic amino acids, which polymerise to form a hydrophobic peptide. This peptide will partition to the protocell membrane for biophysical reasons. Any possible function, such as CO_2 fixation, would depend on that location. Conversely, hydrophilic amino acids interact with RNA containing more hydrophilic letters (C and U). This would form hydrophilic peptides that stay in the aqueous cytosol and potentially interact with metal ions such as Mg^{2+}. Any functions, such as RNA polymerisation, would then depend on that location. This templating of amino-acid sequence in short peptides depends on the sequence of letters in RNA, giving a genetic code linked directly with function, that can evolve through selection on protocell growth.

the heart of metabolism meant that genes could never replace those core biochemical pathways. Information came into being in the context of growing protocells, and to this day genes recreate the ancestral chemistry of their host cells. Far from inventing metabolism, genes *built* on these deep pathways, ultimately freeing cells from their physical cradle, recreating the same ancestral chemistry in ever more distant places. So it is that cells retain an abiding biochemical link to the planet itself, to the chemistry that first gave rise to protocells in hydrothermal vents. This chemistry has been faithfully regenerated by genes ever since, even as the world changed beyond recognition. It is the innermost sanctum of being. Our being.

How to go forwards

This book is not about the origin of the genetic heredity but the Krebs cycle. So let's leave the dense thickets between replicating protocells and the origins of the genetic code for another day, another book. Right now, there are some broad themes that we need to pull out for our longer journey. One critical point is that, under the type of conditions we've been talking about, the synthesis of organic molecules from H_2 and CO_2 is favoured thermodynamically. In other words, a mixture of H_2 and CO_2 has a higher overall energy state than cellular biomass, meaning that CO_2 and H_2 will *release* energy when they react together to form biomass. But some molecules are easier to make than others. If carboxylic acids, amino acids and fatty acids all form spontaneously, RNA synthesis is more difficult. The energy currency in cells that enables these difficult steps to happen is ATP. There's an apparent catch-22 here. ATP synthesis is powered by proton gradients, which rotate the ATP synthase motor, as noted in Chapter 1. This awesome

nanomotor is unquestionably a product of genetic information and natural selection; it didn't just pop into existence by
accident in a prebiotic world. So if the first cells needed ATP
to make genes and proteins, how did they manage without an
ATP synthase?

The answer turns out to be quite simple. ATP can be formed
directly from acetyl phosphate, which in turn is formed from
acetyl CoA, or simpler prebiotic equivalents. Going back to the
reaction steps I outlined earlier, recall that I finished by depicting an acetyl group (as in acetyl CoA), bound to the FeS surface.
If this were to react with an inorganic phosphate bound to the
surface (rather than the oxygen I showed before), the product
released would be acetyl phosphate:

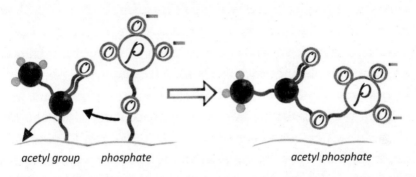

acetyl group   phosphate   acetyl phosphate

Acetyl phosphate remains a critical intermediate between
acetyl CoA and ATP in bacteria and archaea today, as first discovered by Fritz Lipmann. This simple chemistry takes place in
water. Theory says it should occur under prebiotic conditions
too, and we have shown that it really does – acetyl phosphate
is formed from a two-carbon precursor thioacetate. Once
formed, acetyl phosphate will pass its phosphate onto ADP
(adenosine diphosphate) to form ATP at modest (20 per cent)
yield, roughly like this:

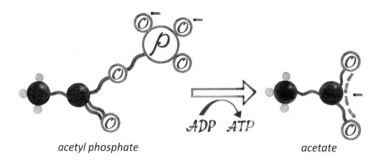

acetyl phosphate   *acetate*

Silvana Pinna, a PhD student in my lab, has shown that ATP synthesis is catalysed beautifully by ferric iron (Fe^{3+}), yet no other metal ions she tried could do the same job. This finding might explain why ATP is the universal energy currency. But the key point is that ATP can be formed directly through simple chemistry in water. Equally critical: this simple chemistry does not release cells from their dependence on proton gradients. On the contrary, these are still needed to drive the synthesis of reactive molecules such as thioacetate and acetyl phosphate in the first place. You can think of the proton gradient as a ratchet, pushing the chemistry of protocells far from equilibrium. In this case, acetyl phosphate and ATP can form readily in locally acidic conditions close to the membrane, but become much less stable (and so more reactive, or further from equilibrium) under more alkaline conditions inside the cell. In terms of reactivity, the difference is about three orders of magnitude, so it's not to be sneezed at.

The bigger picture that emerges is this. Vents provide a steady supply of H_2 and CO_2, in just the right conditions needed to promote their reaction to make carboxylic acids. These form through chemical mechanisms that resemble steps of the reverse Krebs cycle, implying that this chemistry really is the primordial basis for metabolism. The Krebs-cycle intermediates are universal precursors for making amino acids, fatty acids, sugars and eventually nucleotides. Exactly how all those are formed under prebiotic conditions (which catalysts, which feedbacks) is a question of active research in my own lab and others. But some principles are clear. Catalysts and proton gradients do no more than lower the kinetic barriers to reactions that are thermodynamically favoured. How far forwards those reactions can go depends on the driving force, which is the reactivity between H_2 and CO_2. The more H_2 available, the more organic products will be produced, and the further these reactions will be driven from the starting point. You could think of this as gentle waves spreading up a beach; how far up the water travels varies locally with the lie of the land, but ultimately depends on the driving force – the strength of the tide.

The beauty of vents is that hydrothermal flow is continuous. If only a tiny proportion of H_2 reacts at any one moment, in the next moment that H_2 is replenished, and the same reaction can happen all over again. Growth is the non-stop formation of new organics from the continuous flow and reaction of gases in the environment. The inorganic pores within vents are the perfect homes for protocells that nestle in the crannies and tap the local flow, growing, dividing, nestling, growing. Growing.

Let's leave this chapter with a pregnant thought. I have described an 'autotrophic' origin of life, in which gases react on mineral surfaces to form organic molecules – all in a single setting that drives growth over time, though this could be repeated simultaneously in millions of vents spread across the entire sea floor of a water world. The deep structure of metabolism reflects this starting point – CO_2 and H_2 with proton gradients and iron-sulfide catalysts. Metabolism is driven forwards not by itself, but by the environment – ultimately, by the pressure of hydrogen. Practically all the pathways of biochemistry are close to equilibrium, which means that metabolic flux can go either way. The direction of flux is not intrinsic to bio-chemistry but is necessarily extrinsic, depending on the driving force of the environment. Life needs a far-from-equilibrium environment, in which the forward reactions of metabolism are sustained by continuously reactive surroundings, most simply by continuous flow. So growth can continue for as long as hydrogen is available to push metabolism forwards, whether that hydrogen bubbles from the ground, or is prised from the tight grip of water by the power of the sun in photosynthesis. The question is: what happens if the supply of hydrogen dries up? If a vent dies, or hydrothermal flow diverts elsewhere, or later on, if oxygen levels rise in the air? Then the same meta-bolic pathways inexorably go into reverse, oxidising organics to form CO_2 and hydrogen – hydrogen that now burns in oxygen, every atom giving a little kick of energy. The Krebs cycle swings into reverse. And the world turns.

4

REVOLUTIONS

You are not completely spineless. You have a notochord: a flexible rod made of cartilage, which in your descendants, millions of generations hence, will develop into a proper backbone. For now, you flex your rod like an eel to undulate through the water, never quite fast enough. Better to stay submerged in the soft mud at the bottom, with only your head visible, while you filter out grains of food from the swell. You have a wormlike head, with a small bulging of nerves that will one day become a brain. Your eyes aren't much use, but at least you can make out the looming of a monster, and swiftly bury your head again. Oh, times have changed! Not long ago, the world was full of gentle filter feeders, swaying their fronds softly in unison, never harming a soul. Not that you remember, except in some hazy instinctive yearning for the garden of Ediacara. But now there are vast armour-plated war machines, bristling with claws and spikes and rows upon rows of crystalline eye facets fixing you from every direction. You are a tender morsel, barely a couple of inches long, protein-rich muscle strapped to a crispy rod; a tasty snack for *Anomalocaris*. Better pull in your head again, just in case – being a little bit spineless might help you survive in this fearsome new world, outnumbered a thousand to one by spiny monsters.

Well, you did survive. We vertebrates salute you and celebrate your name: *Pikaia*.[1] Later in this chapter, we'll see that hiding in the mud was a wise choice. But how did this revolutionary world of predators and prey come to pass? We are in the thick of the Cambrian period, perhaps 520 million years ago. This is the first time in the four-billion-year history of life that we would recognise our own world. These are creatures with eyes and shells, legs and antennae, behaving in a way that we understand only too well: scuttling for safety or divebombing in attack. These are creatures that live and breathe and die violently, their breath sucked from them. These are creatures that could only exist in an oxygenated world, where there's energy to burn. If we could go back in time and take a muscle biopsy from *Hallucigenia*, I would be fairly certain about one thing: it would have a full oxidative Krebs cycle, spinning like our own. And I would wager a small bet that some Cambrian animals that survived to a mature age would suffer from cancer and other degenerative diseases we know today, the scleroses of old age itself.

How can we possibly know anything about the biochemical workings of creatures that died half a billion years ago, leaving nothing but their hard shells as evidence? The mystery

---

1 Like many people of my generation, I first came across *Pikaia* in an inspiring, if flawed, book by Steven Jay Gould, *Wonderful Life*. *Pikaia*'s standing on the evolutionary path to vertebrates has been queried – for example, it seems to have a cuticle (like invertebrates) – and defended. I am not a palaeontologist and won't offer a view. But culture is important in science too. The unstarred *Pikaia* symbolises the quest for our own lowliest animal origins, making it culturally important, even if not strictly true. The greater truth is that the first chordates were almost certainly small and wormlike in their appearance. *Pikaia* itself is only known from the Burgess shale in Canada, dating back to 508 million years ago (Mya); older examples of probable chordates have been found in the Maotianshan shales near Yunnan, China, including *Myllokunmingia* (518 Mya), *Haikouichthys* (525 Mya) and *Zhongjianichthys* (530 Mya).

is deepened by their abrupt appearance in the fossil record, in what is called the 'Cambrian explosion', when the first recognisable animals suddenly burst into strata of the same age all around the world. Famously, Darwin could 'give no satisfactory answer' to the absence of fossil animals before the Cambrian in *On the Origin of Species*, a puzzle that became known as 'Darwin's dilemma', which seems to challenge the gradualistic notion of natural selection itself. (Spoiler: it doesn't; merely the meaning of the term 'gradual'.) The discovery of large (~1 metre) fossils in the Ediacaran hills of Australia added nuance to the mystery: dating back tens of millions of years before the Cambrian, these enigmatic fossils are usually interpreted as filter-feeding animals tethered to the spot in deep waters, but there is little consensus on how they relate to modern animals. While most palaeontologists interpret a few of them as bilateral animals (roughly symmetrical down the middle, as we are), there is still enough uncertainty for mavericks to claim they were lichens, growing on rocks on land. What everyone can agree about is that virtually all of them vanished without trace soon before the Cambrian explosion, their gentle way of life replaced by the grotesque Cambrian circus of scissor-handed slayers and their victims.

If the fossils themselves are enigmatic, the context established by geology, physiology and ecology point to part of the answer: oxygen. Remember that O_2 accumulated as a waste product from photosynthesis, and was virtually absent from the atmosphere for life's first two billion years. That's no accident. The chances of life starting on an oxygenated planet is arguably close to zero: hydrogen must react with CO_2 to form organic molecules, but does so very reluctantly if at all in the presence of oxygen, which reacts with H_2 far more avidly than does CO_2. That, in a nutshell, is the tension underlying the

Krebs cycle itself, which turned from the chemistry of H_2 and CO_2, driving life into existence four billion years ago, to the chemistry of 2H and O_2, powering the Cambrian explosion more than three billion years later. It turned, quite literally, from the reverse Krebs cycle, making organics, to our own oxidative cycle, burning them, the revolution that powered the revolution.

Why oxygen is unique

The path along which atmospheric oxygen rose to modern levels was long and tortuous. We will picture it from some of the more spectacular vantage points later in this chapter – the Snowball Earth, the Great Oxidation Event and the end-Permian extinction. For now, suffice to say that there is abundant evidence in the rocks that oxygen rose to nearly modern levels around the time of the Cambrian explosion. Whilst oxygen did not light the fuse, the Cambrian explosion would have been a damp squib without it. Oxygen is unique. The O_2 molecule is a 'free radical', meaning that it has an unpaired electron – actually, two of them – which explains its potentially explosive reactivity, as well as its apparently paradoxical tendency to accumulate in the air up to extraordinary levels – nearly 21 per cent – quite a boon for a reactive gas.

The reason for this balance between stability and reactivity stems from the rules of quantum mechanics, which oblige oxygen to react only with molecules that offer up single electrons, such as rusting iron, and not with molecules that have more stable pairs of electrons. Organic molecules don't usually part with single electrons (they deal almost exclusively with pairs of electrons) hence they do not react easily with oxygen. That's why we don't spontaneously combust, and why oxygen can accumulate in the air to such high levels. But of course organic material will burn if set alight with a spark. A fire is a free-radical chain reaction, in which high-energy intermediates can yank single electrons from organic molecules, allowing them to react directly with oxygen. Respiration is a controlled form of combustion. The energy released is exactly equivalent, but it is released in tiny steps to power the synthesis of ATP. Cell respiration goes to great lengths to extract electrons one at a time from '2H', feeding them individually to the fastidious dragon oxygen (see Chapter 1). All this explains why oxygen is unique: when it reacts, it provides plenty of energy, but it will react only under quite constrained conditions. Other molecules such as nitric oxide can provide just as much energy as oxygen, but they react too fast with too many other things to ever accumulate in the atmosphere. As a result, the energy-history of the Earth is the story of oxygen in the air and oceans. It's sobering to realise that without the quantum rules that govern the predominantly two-electron chemistry of carbon, versus the one-electron behaviour of oxygen, the world that we know and love could not exist.

This is how we know from physiology that the Cambrian explosion was fuelled by nothing other than the complete oxidative Krebs cycle. It's a numbers game. Aerobic respiration operates at about 40 per cent efficiency, so I convert about 40 per cent of the energy content of my lunch into usable energy, such as ATP. In the absence of oxygen (and 'bystander' molecules such as nitrate, which only accumulate when oxygen is available) the maximal energy efficiency is closer to 10 per cent. The efficiency constrains the number of trophic levels possible in a food web, as each trophic level has less energy available to it. This means that the total population that could be supported falls off with each trophic level until the energy remaining is too little to support a population at all. Imagine that the minimum energy requirements to maintain a viable population were 1 per cent of the fixed carbon available. With aerobic respiration, at 40 per cent efficiency, five trophic levels could be sustained before reaching 1 per cent (trophic levels 1 to 5 would have 40, 16, 6.4, 2.6 and 1.02 per cent of the energy available in fixed carbon, respectively). Without oxygen, only two trophic levels are possible before we reach 1 per cent. Of course this is a rule of thumb and the reality would depend on the total amount of fixed carbon available, the size of viable populations, the amount of oxygen, and more. Nonetheless, complex food webs are only likely to exist in a well-oxygenated world. More than anything else, the Cambrian explosion marked the beginning of our modern world, with complex ecosystems, multiple trophic levels, competition between predators and prey, the whole shebang. None of that exists anywhere on Earth without oxygen. And in animals, aerobic respiration always spins the Krebs cycle; it is universal. We all descend from those Cambrian beasts, so we can infer that they used the Krebs cycle, linked to aerobic respiration, too.

But an older idea that rising oxygen *drove* the change (as opposed to permitting it) is far from the mark. The fact that animals and plants each arose on just one occasion in the Earth's long history already hints that developmental constraints, such as a genetic framework, also stood in the way of their evolution. If rising oxygen simply raised the curtains to complexity, how could we explain the 100-million-year delay before land plants arose? And why do we never see multicellular animals composed of aerobic bacterial cells? To really grasp the problems involved (and later in the book, to understand what goes wrong in ageing) we need to take the long view, a view that goes right back to the origin of life, where we left off in the previous chapter. There we talked about the Krebs *line*: a linear pathway that arguably preceded the cycle itself. So why did the Krebs line turn into a cycle in the first place? And how did this cycle flip direction to link with aerobic respiration? How did organisms survive when that happened? Flipping the direction of the Krebs cycle did not obviate the need for Krebs-cycle precursors to make all the other molecular building blocks of cells, from amino acids to sugars, fatty acids and nucleotides. At the least, that had to require clever control of flux, for it is not possible to do everything at once – some patterns of metabolic flux are not compatible with others. The Krebs cycle can't go both ways at once ... can it?

I confess I was befuddled by this question too until a simple thought crossed my mind. Perhaps bacteria and single-celled protists are indeed obliged to do one thing or another: they switch on and off genes in different environmental contexts to focus on either growth or energy generation. Serial processing. But multicellular organisms can resolve the problem through parallel processing. Different tissues specialise to perform distinct tasks, each one with its own fairly simple (or at least non-contradictory) flux through the cycle. The deep tension at the heart of the cycle, the yin and yang of biosynthesis versus energy, is resolved in part by multicellularity itself, in which different organs cooperate to carry out different tasks. I'd like to think that this is the glory of the Cambrian explosion – those animals were the first to balance the yin and yang of the Krebs cycle, not just as part of a wider ecosystem, but within themselves. This is far from trivial genetic regulation, for any failure of metabolic flux to find its true path must compromise life itself – our ability to grow, to repair, to move around, see or think. To be alive. In this chapter, we'll see how that came to be.

Plugging into the planet

Let's cast our minds back to the emergence of life in deep-sea hydrothermal vents. With a steady supply of H_2 and free proton gradients, everything that's needed is already in place. Growth can be continuous, and there are at least no thermodynamic impediments to the emergence of genes and cells. But what happens when cells move out from the vents to less bounteous surroundings? Now they must make do with less H_2, or extract it from other sources such as H_2S. Either way, all of the electrons that cells need to grow still derive from geological sources, ultimately the Earth's mantle, which is rich in electron-dense (technically, reduced) metals such as iron and nickel. In contrast, the atmosphere and hydrosphere are relatively oxidised, being composed mainly of gases such as CO_2 that are comparatively electron-poor, giving them a relatively positive charge. From this point of view the Earth is a giant battery, where the inside is more negative than the outside. The reduced mantle connects to the oxidised atmosphere and hydrosphere through hydrothermal vents and volcanoes, gateways to the infernal interior of our planet. Before the evolution of photosynthesis, the flow of electrons that powered all life on earth emanated through these hellish geysers. The current from this giant battery set an upper limit to the size of the biosphere, which must have enforced stringent economies on nascent life.

The single biggest problem facing the first cells was the need to generate their own proton gradients, to drive the reaction between H_2 and CO_2. The topological structure of cells – reduced and alkaline inside, oxidised and acidic outside – is strictly analogous to a vent, nay, to the Earth itself. Cells are mini-batteries that recapitulate the Earth. We saw in the previous chapter that the flow of protons across membranes (the proton-motive force) can modulate the reactivity, or technically the reduction potential, of H_2 and CO_2, as well as of membrane proteins such as Ech. Overall, the proton-motive force drives the difficult transfer of electrons from H_2 onto the FeS protein ferredoxin, which then readily fixes CO_2. This inward flow of protons is the ancestral state. Apart from ferredoxin, the proton-motive force also drives ATP synthesis. As we've seen, both ATP and ferredoxin are needed to spin the reverse Krebs cycle to drive CO_2 fixation. So, when the first cells escaped the vents, they had to generate their own proton-motive force to drive these same processes. Proton pumping is costly; but worse, it draws on the same hydrogen emanating from exactly the same hellish sources.

The details of how this works needn't concern us.[2] What matters is that H_2 is consumed in generating a proton gradient, which powers the synthesis of the ATP and ferredoxin needed

---

2 But I can't resist telling you anyway. In some of the most important and beautiful work in bioenergetics in the last decade, Rolf Thauer and Wolfgang Buckel finally uncovered a clever sleight of hand whose very name conjures up subterfuge: bifurcation. The two electrons from H_2 are separated deep within the inner sanctum of a large protein riddled with iron–nickel–sulfur clusters. One electron leaps away towards a relatively positive charge, following the normal rules of chemistry, while the other is forced off improbably towards a relatively negative charge, landing on ferredoxin. Energetically, it works because the two electron hops are always linked together – the favourable hop releases just enough energy to drive the unfavourable hop. This sophisticated process of electron bifurcation turns out to be practically ubiquitous in anaerobic bacteria and explains many long-standing puzzles. This is the way, for example, that the non-photosynthetic bacteria living in deep-sea hydrothermal vents, mentioned at the close of Chapter 3, generate the ATP and ferredoxin needed to spin their reverse Krebs cycle. The key point to appreciate about electron bifurcation, however, is that it does no more than provide the power to pump protons across a membrane. It drives CO_2 fixation *indirectly*, through the intermediary of the proton-motive force, as described in the main text. Afficionados may note that many of these cells pump sodium ions rather than protons; but their growth depends on a sodium–proton antiporter, so converting a sodium gradient into a proton gradient seems to be necessary.

for fixing CO_2. For many bacteria, even proton pumping requires ferredoxin. The enzyme captures electrons from H_2 and passes them onto an electron acceptor; the energy released is used to pump protons, just as in respiration. On the early Earth, the most common electron acceptor was not oxygen, but most likely CO_2 itself. Transferring 2H onto CO_2 – repeatedly – generates methane (CH_4) and water (H_2O) as waste products. In this case, each 2H transferred can pump two protons, just a fifth of the energy available from aerobic respiration. Even so, repeating this same process over and over again does generate a proton-motive force sufficient to drive CO_2 fixation to generate biomass. The cells that live this way, most notably methanogens, make 40 times as much waste, by mass, as new cell biomass. In other words, of all the H_2 that a methanogen consumes, only a fortieth is converted into biomass and the rest is used to power pumping. That's a not a trivial expense, as the fitness of a cell is best measured in its progeny – how many copies it can make of itself, which is to say, how well it can grow and reproduce. Spending about 98 per cent of a restricted energy budget on charging the membrane, rather than growth, seems excessive, but that's exactly what these ancient cells do. (We will come to another reason why they might need to pump in the Epilogue.) All this puts a premium on every last drop of hydrogen, and especially that most valuable biological currency, ferredoxin.

We keep coming round to ferredoxin, the red protein, so central to all metabolism. Its deep antiquity and central role in life was most clearly appreciated by the great pioneer of bioinformatics, Margaret Dayhoff. The very same year that Daniel Arnon and colleagues first reported the reverse Krebs cycle, 1966, Margaret Dayhoff (working with Richard Eck) published a paper that launched a thousand ships in the journal *Science*. Dayhoff had done her degree in maths at New York University before taking a PhD in quantum chemistry at Columbia, using punched cards to calculate resonance energies of chemical bonds. She joined the National Biomedical Research Foundation as Associate Director, where she collaborated with the celebrated cosmologist Carl Sagan, synergising her work on bond energies with her computational skills to develop programmes that could calculate the equilibrium concentration of gases in planetary atmospheres, including Venus, Jupiter and Mars, as well as the primordial atmosphere of the Earth. I'm struck that Carl Sagan was at that time married to Lynn Margulis; whatever else he did, perhaps his greatest bequest to humanity was kindling the cosmological perspective of two of the most brilliant women of the twentieth century. Uniquely, Dayhoff could link a deep understanding of the quantum mechanisms underpinning planetary processes such as photosynthesis (grounded in bond energies) with her pioneering computational work on comparing protein sequences – the basis of modern phylogenetics – to reconstruct the history of life on Earth like never before. The beginning of it all was her classic paper on ferredoxin in 1966.

The idea that the details of evolutionary history could be written in the slight differences in amino acid sequence between proteins goes back to Francis Crick in 1958, and was first applied in the early 1960s by Linus Pauling and Emil Zuckerkandl. They had considered the differences between the haemoglobin proteins of horses, gorillas and humans – recent history. Dayhoff was far more ambitious. Developing a single-letter code for amino acids that is still used today, she put forward a series of physical chemistry arguments for the extreme antiquity of ferredoxin. She then showed that the sequence of amino acids was roughly repeated, suggesting that the original protein had been duplicated. Next, she showed that there is a repeating sequence of four of the most ancient amino acids – punctuated by occasional later additions – which she argued was the original sequence of ferredoxin, going back to a time before the genetic code had incorporated all twenty amino acids. Before the code was completed! This was amazing, revolutionary stuff, and I still get a thrill from reading it – how to build a protein from scratch; how to tease out patterns hidden in protein sequences to reveal evolutionary history; how to link primordial chemistry with function through the deep conservation of sequence. This is standard stuff today, but seems to have leapt, fully formed, from the few pages of her paper in *Science*.

As more bacterial proteins were sequenced, Dayhoff was able to use ferredoxin and other bioenergetic proteins such as cytochrome c to lay out the history of life. The most ancient cells, her trees indicated, were bacteria such as *Clostridium*, some of which live from H_2 and CO_2 through exactly the mechanisms we have discussed here (in much the same way as methanogens). She also called attention to the antiquity of ancient forms of photosynthesis in bacteria such as *Chromatium* and *Chlorobium*, which use H_2S as an electron donor rather than water. This trick freed them from the onerous cost of using scant H_2 to pump protons instead of fixing CO_2, which was faced by non-photosynthetic bacteria. In *anoxygenic* photosynthesis, chlorophyll is used to strip electrons from H_2S (much easier than stripping them from water) which are then passed onto ferredoxin directly. The waste product is not oxygen but sulfur. The huge advantage here is that the sun now powers the transfer of electrons from a donor (H_2S) to an acceptor (ferredoxin) and then onto CO_2, without the need for burning fuel to power pumping – the 40-fold cost for methanogens. The sun provides this power instead. But there is still a serious disadvantage compared with the more familiar *oxygenic* photosynthesis (where oxygen is the waste), for these bacteria are still plugged into the planet, in that all their electrons derive from geological sources such as volcanoes and hydrothermal vents.

Anoxygenic photosynthesis suffers from a more subtle problem too: it can *either* make ATP *or* reduce ferredoxin, but it can't do both at the same time. That's because bacteria capable of anoxygenic photosynthesis have a single photosystem, which is capable of either ATP synthesis (via a cyclic flow of electrons from chlorophyll back to chlorophyll, with the electrons energised by light) or CO_2 fixation, by transferring electrons from a donor such as hydrogen sulfide (H_2S) onto ferredoxin. Combining both processes together requires wiring the two photosystems in series, which is what happened in oxygenic photosynthesis (we'll come onto this later). The bottom line is that the early photosynthesisers found only partial freedom from the costs of pumping.

Dayhoff went on to show that oxygenic photosynthesis arose before two billion years ago, albeit probably *after* the evolution of respiration, as had been suggested by Otto Warburg decades earlier. Presumably these ancient forms of respiration either used traces of oxygen generated by physical processes such as ultraviolet radiation splitting water, or alternative molecules such as nitrogen oxides (which can be formed by lightning strikes and volcanoes) in place of oxygen. Perhaps most tellingly of all, Dayhoff used the sequence of ferredoxin from plant chloroplasts and cytochrome c from animal mitochondria to show that Lynn Margulis was indeed correct that mitochondria and chloroplasts were once free-living prokaryotes. Sadly, the year after piecing together this detailed tree of cellular life, in 1983, Margaret Dayhoff died of a heart attack at the age of 57. Not everything she wrote about early evolution turned out to be correct; but perhaps the greatest tribute to her importance in biology is that the phylogenetic methods that have now given us a nuanced view of life's early history trace their roots back to Dayhoff's pioneering work.

But I'm getting ahead of myself. The point I want to make here is that before the evolution of oxygenic photosynthesis, the living world was extremely constrained energetically, dominated by the availability of gases such as H_2 and H_2S emanating from volcanoes or hydrothermal systems, which were needed to drive the reduction of ferredoxin and the synthesis of ATP. This world left its imprint on biochemistry. Before we explore how the evolution of oxygenic photosynthesis opened up new horizons, let's first see how these limits are built into the universal core of biochemistry, for they are surprisingly relevant to the emergence of animals in the Cambrian explosion.

Structure of an energetically limited world

Remember our pregnant thought from the end of the previous chapter: most biochemistry is close to equilibrium, which is to say that metabolic pathways can run in either direction. The reason flux goes one way rather than the other reflects the environmental driving force. In hydrothermal vents, that driving force is hydrogen, which pushes flux in the direction of making new organic molecules. But if hydrothermal flow fluctuates, the concentration of H_2 is bound to fall. Leave the vent and the driving force begins to push the other way. Metabolic flux will go into reverse. The main constituents of cells, mostly amino acids and nucleotides, begin to oxidise, draining back down through the same metabolic pathways, potentially offering a new source of energy. As Bill Martin has argued, this was probably the origin of heterotrophy – eating other cells for sustenance. It's also an argument against the old idea of a primordial soup as the origin of life. Any soup is cooked up from hundreds, perhaps thousands, of disparate molecules, each one requiring its own pathway to break down and extract energy. That's not what we see in the living world. We see just a few trunk routes, amounting to the same autotrophic pathways in reverse. This pattern suggests that autotrophic pathways arose first, in environments with a strong driving force, while heterotrophic pathways started out by reversing these pathways in environments where the driving force faltered.

This reversibility of metabolic flux applies to ecosystems rather than single cells. If you are a single cell, the last thing you need is for your metabolic pathways to keep switching direction at the whim of the environment. The first thing you'd want to do is get with the flow; ideally live in a vent, a volcano, or wherever you can maximise the forward driving force. Location, location, location. A second thing you would want to do is cut your costs – minimise the ATP requirements for fixing CO_2. Make the forward reaction as easy as possible. In this respect, it's striking that the ATP costs to make a single molecule of pyruvate via the reverse Krebs cycle are less than half those of the Calvin–Benson cycle (which with oxygenic photosynthesis doesn't face the same energy constraints). But this efficiency only works under optimal conditions, where oxygen levels are close to zero, and so isn't easily divorced from location.

But you can't always be lucky, punk. Sometimes you'll get washed up in the wrong place. You'd better also learn to control your metabolic flux. Accordingly, selection has fashioned key regulatory enzymes to work in only one direction, preventing 'reflux' in much the same way as a valve. Plainly, such one-way systems improve efficiency. Autotrophic cells fix flux in one direction while heterotrophs fix their main pathways in the opposite direction. Bacteria frequently grow from each other's waste products, living cheek by jowl in stratified layers. We've seen that methanogens release methane as a waste product. Methanotrophs, in contrast, gain their energy from oxidising methane, so they benefit from living next to a source of methane. In fact, that benefits both parties – another ecosystem rule. Removing your waste will favour the forwards direction of a metabolic pathway, for the same reason that the assembly line of a factory clogs up if the end-product is not quickly sold on. The most familiar biological example is alcoholic fermentation. Ethanol is a waste product. If it accumulates to more than about 15 per cent, then fermentation grinds to a halt; the pathway is inhibited by its own end-product. That's why wines have no more than about 15 per cent alcohol (you need to distil the alcohol to make spirits or fortified wines). But remove the ethanol and fermentation can start up again. Living close to other cells that devour your ethanol will do that best. The bottom line is that cells contrive to increase the concentration of their substrates, while swiftly getting rid of unwanted waste products, which together pull their metabolic pathways in a forwards direction.

I had a revelatory insight into how far this principle dominates microbial ecosystems a few years ago, when visiting Shawn McGlynn, then working in Victoria Orphan's lab at Caltech. He was using a clever technique called nano-SIMS to image the precise location, within a muddy methane seep (where methane bubbles out from the seafloor), of specific bacteria and archaea tagged with fluorescent markers according to their phylogeny. In the darkened room, he showed me a magical constellation of reds, purples, golds and greens. On closer inspection, the points themselves were multifaceted, composed of clumps of cells, often displaying two different colours, each one marking a different type of bacteria or archaea. The clumps were usually

similar in size, with a comparable distribution of cells, even consistent numbers – a precise stoichiometry. This stoichiometry
varied with the cells' colours, reflecting an intimate metabolic
relationship between the cells: how much waste product was
produced by one set of cells and consumed by the other, how
far gases diffused (or electrons hopped) across the clumps, and
no doubt other unknown subtleties. But whatever the subtleties, they must have been governed by simple and reproducible
laws, as the same patterns always held. Rarely have I felt the
thrill of science so palpably. Here was an extraordinary 'sci-
fi' technology, revealing an unsuspected ecological order to the
bacterial world. At that point they hadn't published their findings, so this was an exciting glimpse into nature's hidden order
that few others on the planet yet knew (it came out in *Nature* a
year later). The lesson I took away was simple: cells rarely live
alone, but collaborate in the most intimate ways to optimise the
driving forces of each other's metabolism. The precise stoichiometry speaks of optimisation, of close collaboration between
genetically distinct cells, a tight 'group hug', in which scarce
resources are traded to keep the flow of metabolism alive. Even
bacteria benefit from the company of others – and this is one
reason why it has proved so difficult to culture most bacteria in
isolation.

I suspect that this reasoning applies to the Krebs cycle itself,
which the biochemist Erich Gnaiger has compared to that
ancient Egyptian (and later alchemical) symbol, the Oroboros,
a serpent or dragon consuming its own tail.[3] This mythical beast

3 Erich Gnaiger is a larger-than-life character, practically a legend in the world
of respirometry, who is steeped in the biophysical tradition of Schrödinger and
Mitchell, and who formed his own company producing high-resolution flu-
orespirometers (descendants of the manometers used by Warburg and Krebs)
called Oroboros. A lover of art as well as science (and schnapps, music and

is usually interpreted as a symbol of eternal cyclic renewal, the cycle of life, death and rebirth. Gnaiger sees thermodynamic meaning too. The Oroboros cycle operates at 100 per cent efficiency: the energy for the serpent's rebirth is captured entirely by consuming its own tail, with no external supply of energy. Life is plainly not a *perpetuum mobile*, so this conception of the Oroboros holds no direct meaning for the Krebs cycle, beyond both of them being cyclic. But it does speak to thermodynamic efficiency nonetheless. We have been talking about the driving force for metabolic pathways. We've seen that optimising the driving force means raising the levels of the substrates that enter the pathway while removing the end products. So the direction of a reaction that is close to equilibrium

can be pushed in the forward direction if the concentration of A is increased, by living next to a good source of it (a vent will do), and B is decreased, by removing it to form C:

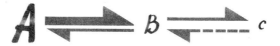

Until now we have considered the Krebs *line* and posed the question: why did it first become a cycle? To sustain forwards flux down the Krebs line, we need to have high concentrations of early intermediates and lower concentrations of the later products. More specifically, we need to be rid of the final product, citrate, to prevent flux oscillating back and forth through the

philosophy), Gnaiger has a wonderful collection of Oroboroi in his gallery in Innsbruck.

pathway. That's what is achieved when the Krebs line is converted into a complete cycle: the end product citrate doesn't build up, but is removed by converting it into its own precursors such as acetyl CoA, the very things needed for more CO_2 fixation.[4] What's more, the enzyme that catalyses this step (ATP citrate lyase) is effectively irreversible, as it consumes ATP to split citrate. So the reverse Krebs cycle really is an Oroboros: by swallowing its own tail, the cycle is continually pulling the forward reactions onwards. For me, the reverse Krebs cycle is an ethereal Oroboros at the very heart of life. It is no solid thing like a serpent, but a fleeting, invisible flux of molecules, still whirling in a cycle after billions of years, sustained by the same atoms that fuel the eternity of stars, hydrogen.

The dawn of photosynthesis

Set against the tight constraints of this anaerobic world, imagine the freedom offered by that everyday miracle, oxygenic photosynthesis. Silently humming in trees and algae and cyanobacteria, all use virtually the same machinery to fix CO_2 and generate ATP. Oxygen is the waste product, cast away, the first great pollutant. Please remember that oxygen does not come

---

4 The eagle-eyed will notice that statement is slightly ambiguous. Citrate is broken down into acetyl CoA plus oxaloacetate, both of which can regenerate citrate. The trouble here is that high levels of oxaloacetate could interfere with the synthesis of more oxaloacetate from acetyl CoA. It therefore seems likely that the reverse Krebs cycle generally regenerates oxaloacetate, whereas the acetyl CoA is bled off to form fatty acids, sugars (via pyruvate) and ATP (via acetyl phosphate). This view could explain how sugar and fat metabolism became separated from the canonical Krebs cycle. If so, then the argument above holds for oxaloacetate rather than acetyl CoA, with the reverse Krebs cycle regenerating oxaloacetate plus acetyl CoA as a high-energy feedstock for most biosynthesis. In any case, the irreversibility of the ATP citrate lyase forces a forward direction of flux.

from CO_2 – it comes from water, which is split apart to extract the hydrogen – 2H. That should sound familiar by now. It's not easy to split water, hence the energy of the sun, focused to work by that marvellous transducer chlorophyll. Literally a transducer, chlorophyll absorbs a photon of light, red light, which excites an electron. The excited electron, zapped away from its former owner, is swiftly spirited off down an electron-transport chain embedded in the membrane. Transiently angry, chlorophyll snatches a replacement from the sacrificial gift offered by the trembling priestly protein. This sacrifice is not some virgin to placate a medieval dragon, but water. Zap. Repeat. Zap. Repeat. It really can do this all day. Electrons flow. Chlorophyll transduces light into electricity. For the first time, life is freed from the churning bowels of our planet. No more must cells rely on volcanic gases, serpentinising seafloor or metal sulfides billowing from black smokers. The oceans themselves have turned into fuel. That distant thermonuclear reactor, the sun, ignites the fuel.

So where do all these electrons go? Ultimately the electrons and protons – the 2H – are united on CO_2, to form organic molecules. We can do better than that: the Calvin–Benson cycle cobbles them onto the C3 molecule phosphoglycerate, to generate the C3 sugar glyceraldehyde phosphate (you can double-check the figure on page 95). The pathway to sugar is both wild and familiar. Wild because the electrons zigzag their way backwards and forwards across the membrane, by way of a second zapping, to … the familiar ferredoxin. And familiar, too, because the electron-transport chain used in photosynthesis is basically the same as the respiratory machinery used by anaerobic bacteria, right down to the same iron–sulfur clusters, which are now repurposed to steal electrons from chlorophyll. These proteins don't lamely transfer electrons to ferredoxin:

they pump protons too, in just the same way as anaerobic bacteria, or for that matter our own mitochondria. And these protons flow back through the same ATP synthase, making ATP. Even the chlorophyll is similar to that used in anoxygenic photosynthesis. The only thing that changed (in a big way) with oxygenic photosynthesis was the overall sequence: the linking of the two existing photosystems in series to give a wild zigzagging Z scheme. And of course the cyanobacteria appropriated the Calvin–Benson cycle in place of the reverse Krebs cycle, at least partly because the former works better in the presence of oxygen, the waste product of oxygenic photosynthesis, only now a global problem.

This is another Earth-changing instance where small steps produce a giant leap. Think about what has changed here. No longer must cells use a small pool of electrons extracted from H_2, H_2S or Fe^{2+} (all derived from hydrothermal processes) to power pumping. And no longer must the proton gradient be used to *either* reduce ferredoxin *or* synthesise ATP, the cramped choice of the anaerobic world. By linking the two distinct photosystems used in earlier forms of photosynthesis into the Z scheme, oxygenic photosynthesis could now generate ATP and reduce ferredoxin simultaneously. Now, the power of the sun sets electrons flowing from water – water everywhere! – to synthesise ATP in the first photosystem, and then reduce ferredoxin in the second. By transferring electrons from water right through the two photosystems linked in series (through the Z scheme) oxygenic photosynthesis frees life from its hydrothermal roots. The tight energy constraints of the anaerobic world finally lift, and life can expand from an invisible film clinging to vents to embrace the planet, now transformed into a blue-green marble that shouts it is alive from space.

Yet that expansion took two billion years, an unfathomable

time. Oxygenic photosynthesis first arose in cyanobacteria or their predecessors, but exactly when remains uncertain. The first unequivocal evidence is the Great Oxidation Event (familiarly called the 'GOE') around 2.3 billion years ago, when the planet turned rusty red and froze. We can attribute both these changes to oxygen, beginning to accumulate in the air. The rusty rocks are mostly composed of iron oxides – red beds and banded iron formations – but other metals became oxidised too, leaching from rocks and accumulating in the environment in colossal ore deposits such as the great Kalahari manganese fields in South Africa, covering an area of several hundred square kilometres, and dating back to 2.2 billion years ago. The global freeze was triggered by oxygen too. Methane reacts with oxygen (either spontaneously or through the auspices of methanotrophs) and methane is a greenhouse gas. Strip it from the air and the planet cools. As oxygen levels rose, more methane was oxidised. So the fine balance tipped, and the Earth plunged into a global glaciation, the first Snowball Earth, locking the whole planet in ice for tens of millions of years, until volcanic emanations of CO_2 eventually warmed it up again.

Setting aside the drama of the GOE and the Snowball Earth, reconstructing the history of photosynthesis is a dastardly business. Just how long before the GOE did the first photosynthetic bacteria arise? Molecular clocks based on gene sequences are fraught with difficulties, as there are few calibration points (such as microfossils that everybody can agree are cyanobacteria) and few constraints on the speed at which genes evolved in deep time. There are purportedly 'whiffs' of oxygen (slightly oxidised minerals) which hint at oxygenic photosynthesis from as long as three billion years ago, but these are bitterly contested and may reflect bacterial metabolism. By the same token, there is next to no evidence against an early evolution of photosynthesis. If all the oxygen being formed reacted swiftly with methane, there would be no trace of oxygen in ancient rocks. I admit that doesn't sound plausible, but common sense is the worst guide to the history of life. In fact, the levels of atmospheric oxygen depend on exactly such a balance – the rate oxygen is produced by photosynthesis minus the rate it is consumed by respiration, rotting, oxidation of minerals and so on. Lumping them all together as 'respiration', these processes balance photosynthesis almost exactly (where 'CH_2O' signifies all forms of organic matter):

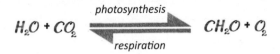

$$H_2O + CO_2 \xrightleftharpoons[\text{respiration}]{\text{photosynthesis}} CH_2O + O_2$$

The long-term balance in these rates explains why the composition of the atmosphere remains nearly constant, at least for the major gases nitrogen and oxygen. But plainly that is not true over geological time – oxygen levels increased from virtually zero to nearly 21 per cent today. There must have been periods of serious mismatch between the rates at which oxygen is generated by photosynthesis and its removal through oxidative processes. For oxygen to accumulate over the geological aeons, the organic carbon formed by photosynthesis has to be shielded from oxidation, for example through burial as coal or oil. The vast majority of buried carbon is actually sequestered in carbon-rich shales, not economically viable to extract, for which we can only be grateful. In principle, knowing how much organic carbon was buried, and when, would give an approximate timeline for the accumulation of oxygen over geological time. In practice, that is hard to know with any certainty. Luckily, there's a get-out-of-jail-free card that gives us a proxy: carbon isotopes. We met two of these previously, in Chapter 2:

^{11}C and ^{14}C. Here, we're concerned with a third isotope, ^{13}C, which happens to be stable. The nucleus of ^{13}C has six protons and seven neutrons. The fact that it's stable means that the ^{12}C to ^{13}C ratio in the environment does not change over time, or at least not as a result of radioactive decay; but ^{13}C is not common, barely a hundredth as abundant of ^{12}C.

Remember rubisco, the world's most abundant protein, responsible for fixing CO_2 in the Calvin–Benson cycle. Rubisco has a slight bias towards the lighter of the two stable forms of carbon, ^{12}C. I think of these molecules as ping-pong balls, bouncing around and occasionally bumping into enzymes such as rubisco. Lighter molecules bounce around a touch faster, so encounter enzymes more frequently. In the case of rubisco, the outcome is that organic carbon becomes slightly enriched in ^{12}C over ^{13}C (compared with a standard ratio) by around 30 per mille (30/1,000). This difference can be measured in modern plants as well as in traces of organic matter trapped in ancient rocks. The bias leaves behind a little more ^{13}C in the oceans and atmosphere. That slight rise is in turn reflected in the composition of limestone, formed by the precipitation of inorganic carbonate from the oceans. Like sea salt, the concentration of carbonate is roughly constant across the oceans. Variations over time in the ^{13}C content of limestone therefore give a global indication of how much organic carbon got buried – the more ^{12}C buried, the greater the ^{13}C abundance in limestone sedimented at the same time.

The soaring ^{13}C peak in the GOE therefore speaks of carbon burial and rising oxygen levels. Converting this trace into an estimate of atmospheric composition is tricky and we don't need to go there; we can just note that high ^{13}C corresponds to the many other signs of oxygen, from red beds to manganese fields. But then the ^{13}C trace flat-lines for around a billion years, sometimes called the 'boring billion', even though that is when complex ('eukaryotic') cells evolved, along with traits such as sex, which are biologically anything but boring. I've dwelt on that theme in other books, but it is less relevant here.[5] Two things matter right now. The first is that atmospheric oxygen remained low throughout the boring billion, at around 1–10 per cent of present atmospheric levels, with the deep oceans remaining anoxic. The second is what happened at the end of the boring billion: 200 million years of planetary convulsions, encompassing a succession of Snowball Earths, and accompanied by extreme variations in carbonate ^{13}C values. Some shot up, testifying to a world flooding with oxygen, but there were cataclysmic downturns too that are harder to interpret. And hard on the heels of the biggest downturn of all came the Cambrian explosion. Understanding this enigmatic downturn is key to understanding the origin of animals.

---

5 But here is a taster, in light of the driving forces we have been discussing in this chapter. There is a startling difference in the NAD^+ to NADH ratio within the mitochondria compared with the cytosol, which was brought to my attention by Wayne Willis at the University of Arizona. This is possible only in eukaryotes, because only eukaryotes have mitochondria as a separate compartment. The mitochondrial inner membrane is not directly permeable to NAD^+ or NADH. When the first steps of glucose breakdown (glycolysis) take place in the cytosol, lots of spare NAD^+ is needed to accept electrons, enabling extremely fast ATP synthesis through glycolysis. The ratio of NAD^+ to NADH is therefore kept at about 1,000:1. In contrast, the optimal ratio within the mitochondria is different, as they oxidise NADH back to NAD^+, hence need enough NADH to feed the respiratory chain. Here the ratio of NAD^+ to NADH is normally kept closer to 1:1, which is to say nearly three orders of magnitude lower than the cytosol. The neat trick is that the mitochondrial membrane potential is used to power a pump (the malate–aspartate shuttle, since you ask) that has the overall effect of oxidising NADH in the cytosol and reducing NAD^+ in the mitochondria, optimising the ratio (and so the driving force) in each compartment. So the mitochondria in eukaryotes don't just enable more ATP synthesis, they also optimise the driving forces in a way that is likely to be impossible in bacteria. In this case, bigger really is better.

The Shuram conundrum

Named after the Shuram Formation in Oman, this carbonate-rich sedimentary rock contains the single largest 'negative isotope excursion' in Earth's history. Beginning in strata dated to 560 million years ago, the ^{13}C trace falls abruptly from around 5 per mille above the global standard, to a record 12 per mille below (much lower than the mantle ^{13}C content), recovering slowly over ten million years. Does that sound uninterestingly dry and technical? It's actually shocking! Just think what it means. A fall in ^{13}C relative to ^{12}C implies that far more carbon was oxidised than was buried for millions of years. That in turn implies that oxygen was being consumed, stripped out of the air, producing a massive oxygen deficit. Yet at the end of this period, by the beginning of the Cambrian, the oxygen concentration in the oceans was approaching modern levels, raising the curtain to our fast-paced world of predators and prey. There's even signs of rising oxygen in the atmosphere during the Shuram itself. You might think there's nothing contradictory about that. Why could oxygen levels not fall, then rise again, or perhaps be different in the oceans compared with the atmosphere? The problem is the unparalleled scale of the dip. You can calculate how much oxygen would need to be consumed to pull down the ^{13}C levels so dramatically over 10 million years. The answer is stark – *all of it*. There would be no oxygen in either the atmosphere or the oceans. And don't think the Shuram Formation is an aberration – other geological formations from the same period around the world say the same thing. The conundrum is real.

My colleague at UCL, Graham Shields, thinks he's solved the puzzle. If he's right (and I think he is) then it has a great deal to say about the evolution of animals. To understand, we need to adjust our view in two ways. First, physical burial in the

Earth's crust is not the only way to sequester organic carbon. It can also simply clog up the water column, giving what Shields calls 'peat-bog oceans', in which the organic content dissolved in the water is similar to Scottish marshes – brown water. When oxygen levels are as low as they were in the deep oceans until the late Precambrian, this material could take long ages to decay. Oxidising it en masse would then simultaneously clear the water column and reintroduce a vast amount of $^{12}CO_2$ back into the system, certainly enough to account for the Shuram excursion. But how could all this carbon be oxidised without stripping the oxygen from the atmosphere? The answer, says Shields, lies in geology, not biology. The oxidant was not oxygen at all, or at least not directly, but *sulfate*. Ever since oxygen levels began to rise, from the GOE onwards, evaporitic sulfate deposits gradually accumulated on land. By the time of the Shuram excursion, sulfate had been accumulating on land for the best part of two billion years. How? As the supercontinents came and went, rifting and breaking apart, large land masses became separated by shallow seas, which partially evaporated to form massive deposits of evaporitic sulfate minerals such as gypsum. When the drifting continents brought Gondwana together, around 560 million years ago, these great accumulations of sulfate were raised up high as continents collided and mountain ranges rose. Eroding from the uplifting mountains, a great flux of sulfate flooded back into the oceans. This flood reflected the power of plate tectonics, but also the hand of chance: the happenstance position of large gypsum evaporites in relation to water basins. In short, it could easily never have happened.

Like oxygen, sulfate is an electron acceptor, albeit feeble in comparison. Whilst bacteria first learnt to use sulfate in the ancient anaerobic days before the dawn of photosynthesis, its abundance increased massively in an oxygenated world.

Sulfate-reducing bacteria are heterotrophs, which gain their energy by stripping electrons from organic molecules – food – and passing them on to sulfate, in place of oxygen, producing hydrogen sulfide (rather than water) as a waste product. Some of this sulfide can react with dissolved iron to form insoluble iron sulfides such as iron pyrites, also known as fool's gold. And that settles down to the bottom of the oceans, where it is buried, along with its electrons stripped from dissolved organic carbon. In other words, the peat-bog oceans were oxidised by sulfate, not oxygen, liberating a vast amount of $^{12}CO_2$ back into the system, while burying its long-lost electrons not as fossil fuels but fool's gold. Who's laughing now.[6]

What does this say about the early evolution of animals? It suggests that the oceans were reeking with hydrogen sulfide for the best part of ten million years, until all the organic carbon clogging up the water column had been oxidised. This whole-sale oxidation consumed little if any oxygen, so the world that emerged at the other end was clear, oxygenated waters – perfect for Cambrian predators. But why those animals, and not the gentle swaying fronds of the preceding Ediacaran fauna? The Harvard geologist Andy Knoll has pointed to a clue from another dramatic epoch in Earth's history, the end-Permian extinction, around 250 million years ago. That too was a period of global warming, associated with a toxic mix of falling oxygen,

---

6 Certainly not Shields. He has the bit between his teeth. The Shuram anomaly is not the only conundrum in the carbon isotope record. Several others might well have been confounded by sulfate oxidation too, or even the opposite – sulfate deposition. If the large Shuram negative ^{13}C excursion can be explained by sulfate flooding into the oceans, the large positive excursion immediately after the GOE might have been caused partly by the sequestration of sulfate on land. That could explain why the ^{13}C signal is so large, yet soon afterwards oxygen levels plummeted to nearly zero. The idea that ^{13}C variations only reflect carbon burial and oxygenation is clearly misguided.

rising CO_2 and reeking sulfidic oceans. The oceanic dead zones expanded to asphyxiate a barely credible 95 per cent of species. Yet the Grim Reaper was not undiscriminating. Knoll points out that the 5 per cent survivors were far from random, but primarily included animals with respiratory and circulatory systems that could be actively ventilated, animals that managed to deliver scraps of oxygen to their muscles and clear out excess CO_2 or sulfide. Those tough survivors had already adapted to burrowing through the stagnant muds, remaining active in even the most stifling conditions. The gentle filter feeders that had made it through to the Permian had low oxygen requirements but no circulatory systems. They died.

Three hundred million years before the end-Permian extinction – 550 million years ago – it seems that the Ediacarans, too, were snuffed out in sulfidic seas, unable to breathe. The writing was not on the wall, but in the mud. The first 'trace fossils' (literally, the traces of burrowing animals, like those worm casts you still see on beaches) date to around the time of the Shuram excursion. These were bilaterally symmetrical, muscular, worm-like animals, predecessors of both chordates like *Pikaia* and invertebrates. They had simple circulatory systems, which originated as little more than an open (later closed) body cavity surrounded by muscle, the heart being simply thicker muscle. The early bilaterians already had ways of storing and circulating oxygen, as well as removing CO_2, using pigments such as myoglobin and haemoglobin. They could deal with moderate amounts of sulfide through enzymes that were nearly ubiquitous in these groups (such as sulfide-quinone reductase and the alternative oxidase). Together, these stripped electrons from hydrogen sulfide and passed them ultimately onto oxygen, detoxifying the sulfide in the process. This is an active strategy – not surviving in some passive, metabolically depressed state, but grappling physically with low-oxygen conditions, flourishing with a flickering flame. A thousand thousand slimy things lived on, and so did I, or at least did my ancestor *Pikaia*, slithering through the gloop. Far from oxygen sculpting the cell physiology of animals, these were the conditions that fashioned the Krebs cycle as we know it, with all that means for us today.

The view from inside our cells

We left the Krebs cycle as an Oroboros, consuming its own tail whilst transforming the gases CO_2 and H_2 into the molecules of life. Just how common the full cycle might be among bacteria and archaea remains uncertain, but one thing is clear – most bacteria and archaea are not committed to the full cycle turning in either direction. For much of the time they split the 'cycle' into two halves, a forked pathway rather than a cycle (see the scheme below). One prong of this fork is reductive, the other oxidative. The reductive prong begins with the C2 acetyl CoA and runs through the C3 pyruvate to the C4 oxaloacetate, then onto malate, fumarate and succinate. This is the reverse pathway used in biosynthesis (as anticipated by Szent-Györgyi in the 1930s) that we have followed over the last few chapters. The oxidative prong corresponds to the first few steps of Krebs's own vision, where acetyl CoA instead condenses with oxaloacetate to form the C6 citrate, running on through isocitrate to the C5 α-ketoglutarate (which is used to synthesise the important amino acid glutamine) and sometimes beyond to the C4 succinyl CoA.

So what's the point of prongs, you may wonder? The answer is they bring balance to the force, Padawan – balance to growth. If oxygen levels are low, one prong consumes NADH (shown here as 2H) while the other generates it. Together, they maintain a poise between the donors and acceptors of 2H, while allowing a little ATP synthesis on the side.

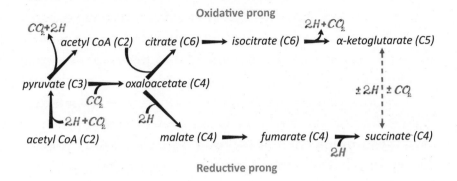

The two prongs of the Krebs non-cycle, a common flux pattern in microbes. The prongs balance the formation of 2H against its consumption when oxygen levels are low, allowing metabolism and ATP synthesis to keep going without a large accumulation of NADH or waste products such as lactate.

The two prongs of this pathway meet in the vicinity of the C4 succinate. In the reverse Krebs cycle, the enzyme that converts fumarate into succinate is called fumarate reductase. This is unique in being the only enzyme in the cycle that is still embedded in the mitochondrial membrane, and it can capture electrons from the respiratory chain. In the absence of oxygen, fumarate can therefore act as a terminal electron acceptor, generating succinate as a waste product, while allowing some proton pumping and ATP synthesis (we'll see how in the next chapter). Accumulation of succinate is a powerful signal for physiological adaptation to low-oxygen (hypoxic) conditions, switching many genes on and off. Conversely, when oxygen is available fumarate reductase is switched off, and an alternative enzyme is expressed instead – succinate dehydrogenase.

Although structurally very similar (clearly derived from a common ancestor) this does exactly the opposite, stripping electrons from succinate to generate fumarate, feeding the electrons directly into the respiratory chain and on to oxygen. In other words, succinate is a tipping point, where Krebs-cycle flux could go one way or the other depending on the external conditions – how much oxygen is available.

The idea that oxygen levels rose swiftly (we've seen they didn't) forcing the Krebs cycle into reverse (or rather, our own forwards direction) is far from the mark in every respect. Oxygen levels rose around the time of the GOE, 2.3 billion years ago, but then remained low (in the deep oceans close to zero) for nearly two billion years. More broadly, the conception of the Krebs cycle as a perfect Platonic cycle, irreducibly complex and unevolvable, is about as wrong as can be. Some groups of bacteria closed the Oroboros for maximum efficiency, but most bacteria, archaea and single-celled protists – let's just stick with the old term microbes – use a two-pronged non-cycle most of the time. Of course, some aerobic microbes can close the oxidative cycle too, but most of them don't use a complete cycle most of the time.

The prongs point to another interesting balance. Microbes are at the mercy of a changing environment. They must switch genes on and off, adjusting their metabolic state to survive. Each metabolic state equates to a particular flux pattern, and so microbes change states in series, over time, from one state to the next. Ironically, very little is known about the real-world metabolic versatility of microbes. We'll consider optimal Krebs-cycle flux for fast growth in cancer cells in the next chapter, as more is known about them, yet much of that came as a revelation. Cancer cells are sometimes claimed to have 'reverted' to a more primitive energy phenotype but that's misleading.

They certainly don't have a 'textbook' Krebs cycle, although not many cells do. The critical point to appreciate now is that while the complete oxidative Krebs cycle might be the most efficient way of generating ATP, it is rarely the optimum to support growth and replication. ATP is just one aspect of growth. The optimisation of biosynthetic pathways to drive growth does not call for the canonical Krebs cycle, but often something quite different.

So why do textbooks still cling to the oxidative Krebs cycle and ATP synthesis? What is the difference between microbes and multicellular organisms? And what happened in that traumatic age before the Cambrian explosion? Crawling around in the mud, snuffling through the sulfide, growing and reproducing in this stench, puts a premium on efficient energy use, as tight as that in the ancient anaerobic world. And the same principles can help to guide us here too.

The first principle is to maximise the driving force – increase the concentration of oxygen, and be rid of CO_2 waste. Ventilate your respiratory system. Store oxygen to be released slowly when needed, which is just what myoglobin and haemoglobin do. The genes encoding both globins were duplicated repeatedly in early animals at this time, giving rise to large families of related proteins, which specialised to slightly different purposes. Next, minimise ATP use, or maximise its synthesis, which is broadly the same thing. For very active tissues such as the brain, that means a complete oxidative Krebs cycle, generating as much NADH as possible, maximising the pumping of protons, regenerating substrates at high concentration. In short: for efficiency, close the cycle. The oxidative cycle is an Oroboros too, with its tail turned into its head. Krebs knew well that the cycle optimises the efficiency of ATP synthesis (it even burns two molecules of water, as we noted earlier). And that means fixing the direction of flux, by way of a few critical enzymes that don't switch direction easily; precisely those enzymes that once seemed to make the reverse cycle impossible (we've seen they don't really).

Yet the complete oxidative cycle, by its very nature, cannot generate the precursors for biosynthesis. To do so requires flux in and out of the cycle, turning it into a roundabout, and perhaps even calling for reverse flux through some parts of the cycle. Of course, biosynthesis still needs ATP. Tissues with much lower oxygen requirements, such as the gonads, ferment sugars via glycolysis, or use unusual forms of respiration to make their ATP (we'll meet some in the next chapter), but that doesn't mean they don't need their mitochondria. On the contrary, their Krebs cycle is running in a biosynthetic mode, not unlike many cancer cells. Different tissues have different needs, some with high ATP demands, others with specific biosynthetic requirements. Each of these states comes with optimal flux patterns through the Krebs cycle.

The multicellular organisation of animals allows tissues to change their metabolic states in *parallel*, with the flux pattern in one tissue balancing that in another. This is the most delicate balance of all, and it could underpin the evolution of complex multicellularity in animals.

We've seen that bacteria and archaea often benefit from company, living symbiotically, with a precise stoichiometry in which one cell type consumes the waste products of another. That helps them to stabilise their metabolic state, creating a metabolic homeostasis that shields them from the vicissitudes of the environment. The multiple tissues of animals could be seen this way too; to some extent, individual tissues maintain a stable metabolic balance by importing or exporting each other's waste products – perhaps exporting lactate or glutamine to be oxidised elsewhere. We're most familiar with the idea of homeostasis maintaining a stable body temperature, pH, salt balance and so on. The notion that organisms maintain an overall homeostasis by stabilising complementary patterns of metabolic flux in different tissues is less familiar, but the large genomes of eukaryotic cells enable exactly this. The first animals, the Ediacaran fauna, had little tissue differentiation, and were doomed when the environmental conditions shifted outside their comfort zone. For them, it was the dead zone. But the burrowing bilaterian ancestors of the vertebrates and invertebrates were already experimenting with a high-wire metabolic balance. By the dawn of the Cambrian they had perfected the art. When the seas cleared to reveal sparkling oxygenated waters, these animals already had a complete Krebs cycle in their muscles and brains, balanced by a range of biosynthetic flux patterns in other tissues. Rising oxygen just gave them a turbocharge.

We saw in Chapter 1 that Krebs had been lucky in his choice of pigeon-breast muscle in conceptualising his cycle. Other tissues exhibit more complex behaviour. Yet the idea of a simple cycle burning glucose persists as textbook knowledge. So it came as a shock to discover that the regulation of the Krebs cycle is far more complex than had been expected. From the beginning, the parallel control of metabolic flux as a symbiosis between tissues was a high-wire act. It could never work without regulatory finesse, and that in itself goes some way to explaining the unique origins of 'organ-grade' complexity in animals and plants, but nothing else. Now the power of metabolomics is being brought to bear on the underlying metabolic basis of age-related diseases, from cancer to diabetes and neurodegeneration. The simplistic notion that genes control metabolism is beginning to unravel, exposing the delicate symbiosis between tissues that underpins the health and lifespan of animals. Hypoxia, infections, inflammation, mutations – all can alter flux patterns through the Krebs cycle, with a knock-on effect that switches on or off hundreds or thousands of genes, changing the stable (epigenetic) state of cells and tissues. Tissue function eventually becomes strained, biosynthetic pathways falter, ATP synthesis declines and the delicate web of symbiosis between tissues begins to fray. And so we age.

5

TO THE DARK SIDE

'The dream of every cell is to become two cells' said François Jacob, the most lyrical revolutionary of molecular biology. No cell lives the dream so wholly or so senselessly as a cancer cell, turning dream to nightmare. Nothing else captures the myopic immediacy of natural selection so starkly. The moment is all that matters for selection: there is no foresight, no balance, no slowing at the prospect of doom. Just the best ploy for the moment, for me, right now, not for the many, and often mistaken. Cancer cells die in piles, necrotic flesh worse than the trenches. The decimated survivors mutate, evolve, adapt, exploit their shifting environment, selfish to the bitter end. Their horror is that they know no bounds. They will eat away at our flesh to fuel their pointless lives and deaths, until, if we are unlucky, they take us too. I am writing about cancer, but must confess that I have the pointless greed and destruction of humanity at the back of my mind. May we find it within ourselves to be better than cancer cells.

Hitler declared war on cancer and lived in mortal fear. So too did President Nixon, in what he hoped was his administration's most significant act. Who would have thought that Nixon's war on cancer, declared in 1971, would be so much harder than landing on the moon, chosen by John F. Kennedy just a decade earlier and already accomplished. The Manhattan Project to make the atomic bomb took but three years. Who knows what we have spent on cancer in comparison, but it is at least ten times as much, with no end in sight. According to the US National Center for Health Statistics, the overall death rate from malignant tumours has barely changed since 1971.[1] Of course there has been progress, and incremental progress goes surprisingly far, especially for early cancers, but most of us have friends or family members who died before their time while we looked on helplessly. We know we don't know the answer.

When I first studied biochemistry in the 1980s, the answer seemed tantalisingly close. A new paradigm was emerging, of oncogenes and tumour-suppressor genes. These are the genes that control the cell cycle, which, when mutated, can drive cells to divide and divide, heedless of the context. Ostensibly random mutations began to make sense in much the same way that scattered dots, when joined up dot to dot, begin to depict a meaningful image. There is a scientific thrill as understanding emerges. What's more, this new paradigm was grounded in the prevailing vision of biology: information. The central dogma of biology, that information flows from DNA to proteins, morphed into the central dogma of cancer, that mutations in DNA produce defective proteins that distort the meaning of signals. In a malevolent form of Chinese whispers, normal

---

1 This may sound surprising, as there has been much greater success in early diagnosis and treatment of cancers. In fact, the overall cancer death rate has declined steadily since 1990, but the longer-term trend shows it rose slowly from 1958 to 1990. The overall cancer death rate in 2016 was barely lower than the rate in the 1950s. My source is the National Center for Health Statistics. Deaths: Final Data for 2016, National Vital Statistics Reports, Vol. 67 No. 5, 26 July 2018 (Figure 6): https://www.cdc.gov/nchs/data/nvsr/nvsr67/nvsr67_05.pdf.

signals are transmuted into the instructions *keep growing!* and *don't die!*

I hardly need to say that this new understanding was developing at a time when it was becoming feasible to sequence genes and pinpoint the mutations involved. Exposing the precise mutation often made it clear exactly how the harm was done. The stars were aligning – the powerful methodology of gene sequencing could link alterations in single DNA letters to changes in the mechanisms of proteins, with clear downstream consequences for signalling and transformation into a cancer cell. Not surprisingly, this combination of conceptual simplicity with practically infinite detail proved irresistible – just point the sequencer and the answers reel out. It even seemed to be objective, as there is no subjective hypothesis to test – just the hard data from the sequencer, constrained by no more than the paradigm itself (which is, of course, a hypothesis). As I write, the Cancer Genome Atlas has catalogued more than three million oncogenic mutations in over 23,000 genes. One might be forgiven for wondering if we have lost the wood for the trees.

The idea that mutations cause cancer remains the dominant paradigm. A special issue of *Nature* from 2020 wrote: 'Cancer is a disease of the genome, caused by a cell's acquisition of somatic mutations in key cancer genes.' Yet over the last decade it has looked as if the juggernaut has rolled too far. It has certainly failed to deliver on its promise in terms of therapies. So why hasn't the death rate from malignant cancer changed since 1971? The oncogene paradigm is not actually wrong, but neither is it the whole truth. Oncogenes and tumour-suppressor genes certainly do mutate, and they certainly can drive cancer, but the context is far more important than the paradigm might imply. We are not immune to dogmas even today, and the idea that cancer is a disease of the genome is too close to dogma.

Biology is not only about information. Just as human delinquency cannot be blamed on individuals only, but partly reflects the society in which we live, so the effects of oncogenes said to cause cancer are not set in stone, but take their meaning from the environment. The idea that 'one rogue cell' causes cancer is a scientific morality tale. The rogue cell is an unfortunate product of multiple blind hits to its genes, acts of arbitrary violence in an uncaring world. By chance (or hereditary misfortune) the miserable cell takes one hit too many and sets out on a rampage of proliferation, unrestrainable by the rational cues of the body politic. That's it: a cold lack of meaning in a cruel world, bleak as the Joker bent on revenge.

Yet the evidence against this dismal genetic determinism is startling. If a cancer is a clone deriving from one rogue cell, then the mutations that caused it to go awry should be written into all its descendants: all the cells in the tumour. That's not always the case. Different mutations are found in different parts of many tumours, often with little if any overlap. This implies that the mutations accumulated during the growth of the tumour, rather than triggering its inception. In fact, later accumulation is quite plausible, as cancer cells are well known to develop genetic instability, meaning that they accrue new mutations much faster than normal cells. These may well exacerbate disease progression, but such heterogeneity is inconsistent with oncogenic mutations *causing* the cancer in the first place. Equally incriminating, the same oncogene mutations are often found in normal tissues surrounding a tumour, or far away elsewhere in the body. Plainly they do not always cause cancerous growth. On the contrary, taking cancer cells from a tumour microenvironment and implanting them in a normal cellular environment tends to halt their growth, and often promotes programmed cell death. It's quite encouraging that cancer cells can be rewired in this way: the same genetic mutations are not invariably associated with the same cell fate. This view is consistent with older work which shows that transplanting the nucleus from one cell type to another does not alter the phenotype of the recipient. The cytoplasm determines the cell state, not the genes in the nucleus. In other words, genes are switched on or off by signals from inside (or outside) the cell. Mutations are not necessary to cause cancer – merely a cytoplasm that sends the wrong signals.

I could go on. Many carcinogens do not cause immediate mutations but may take many years to induce cancer, long after they vanished from the body. Perhaps one-third of cancers world-wide are associated with chronic infections such as hepatitis B and C, and schistosomiasis, which can mutate genes (as they insert into DNA) but also induce cells to proliferate through pulling on the normal levers of cell division. Which of these actually causes cancer is unclear. Perhaps the greatest risk factor for cancer is older age: cancer incidence increases exponentially with age. People over the age of fifty have around *ninety times* the risk of people younger than twenty-four. I'm now in my mid-fifties. My risk of cancer will double by the time I'm in my early sixties, and double again by my seventies. Grim prospects.

You might think this risk profile is explained by the steady accumulation of mutations with age, some of which just tragically happen to be in the wrong genes, causing cancer. Yet the build-up of mutations with age seems to be too slow to explain either cancer or ageing as a process. Nor can it explain why humans do not have a higher cancer rate than, say, mice, despite having ten times as many rounds of DNA copying to make an individual (each of which introduces new mutations), or why elephants have a lower rate still. None of this is to say that mutations don't occur in cancer; obviously they do. Nor am I suggesting that they cannot play a decisive role in its development; they can certainly lock cells into a pattern of growth, making it harder to revert (although we've noted that changing the microenvironment can switch off growth). The question is, are genetic mutations the primary cause of cancer? And if not, what is?

An alternative, and historically controversial, take on the prime cause of cancer has been lurking in the background for nearly a century, going back to Krebs's own mentor, the great biochemist Otto Warburg. It has risen to prominence in the last decade, in part through the unexpected discovery that mutations in genes encoding Krebs-cycle enzymes can cause cancer. Whereas Warburg viewed cancer as the result of damage to the machinery of respiration – an energy problem – this view is not sustained by the myriad of details we now know. But metabolic flux through the Krebs cycle clearly must be central. We've seen that the Krebs cycle is not only about energy, but also about growth. Cancer, first and foremost, is a problem with aberrant growth. While enormously heterogeneous in its manifestations, it seems to me that the key to understanding cancer metabolism is always 'follow the money' In other words: how does any metabolic shift, genetic or otherwise, facilitate *growth* (whether it be to press down the levers that drive growth, or lift those that suppress growth)? In this chapter, we'll see how the tension between energy and growth, the yin and yang of the Krebs cycle, begins to explain the root cause of cancer.

The Warburg effect

Otto Warburg was a genius. He could have won the Nobel prize three times and it would not have been undeserving. In some respects, he was half a century ahead of his time. From the 1930s, he advocated quitting smoking, avoiding pollution from vehicles, eating healthily (he shunned fertilisers), exercising regularly and supplementing one's diet with B group vitamins (which help form NADH, the hydrogen currency in cells, also discovered by Warburg himself). His explanation for cancer has come back into fashion over the last decade, with a near-exponential rise in the number of papers citing the so-called 'Warburg effect' – the tendency of cancer cells to behave like yeast, fermenting glucose rather than respiring, even when oxygen is present. That's to say nothing of his many other discoveries, including much of the other hardware for respiration and fermentation. He was also an extraordinary mentor for numerous other scientists, many of whom went on to win the Nobel prize themselves, including Krebs himself.

Yet Warburg also illustrates the problem with a simplistic notion of right and wrong in science. He was forceful in his views, and his dogmatism meant that he would argue rather than listen, sometimes for decades. His haughtiness was no doubt attributable in part to his extraordinary upbringing in Imperial Germany before the Great War. His father Emil was a distinguished physicist, whose own parents were Orthodox Jews; Emil had apparently converted to Protestantism after arguing with his parents. Otto's mother was from a Protestant family of military and governmental officials from south Germany, and was both sociable and resolute. In 1896, the family moved to Berlin when Emil became Director of the Institute of Physics. As a member of the Prussian Academy of Sciences, Emil became close friends with some of the scientific celebrities of the age. The cancer biologist Angela Otto wrote:

The house of the Warburgs was the site of vibrant social

evenings, where Einstein played the violin, Planck played the piano, and other colleagues such as J. H. van't Hoff and Walter Nernst contributed to the musical, literary and philosophical entertainment. No doubt these guests seeded and fostered Otto's interest in natural science and molded his personality.

It's striking how many great scientists are also culturally rounded people, though I can't help calling to mind a quip from the great violinist Fritz Kreisler to his friend Einstein – the two played together in string quartets: 'You know Albert, your trouble is that you can't count!' Towards the end of the First World War, Einstein wrote to Otto at the behest of his mother, urging him not to return to the Russian front after an injury (for which he was awarded the Iron Cross) but instead to go back to his lab in Berlin. Otto acquiesced. The war was the only break he took from research in his life.[2]

This irresistible company of physicists surely influenced Otto Warburg's vision of biology. While not a physicist himself, his perspective on biology always sought the luminous simplicity of a physicist. Biology is rarely so law-abiding. Warburg sustained at least three prolonged arguments about fundamental aspects of biology. The first was with David Keilin about the nature of respiration. As we noted in Chapter 1, Keilin had identified three cytochromes, which he argued transferred electrons

---

2 Warburg actively enjoyed the war, which reminds me of J. B. S. Haldane, who also relished the First World War. Haldane reputedly took a bet that he could cycle safely across no-man's land, because the Germans would be too surprised to shoot at him – which he duly demonstrated. It would be interesting to know if there is any correlation between those given to bold thinking in science – which unquestionably applied to both Warburg and Haldane – and a tendency to enjoy physical danger.

to oxygen in series, as a respiratory chain. Warburg would have none of it. He saw the need for no more than his own 'respiratory ferment', now known as cytochrome oxidase – a single reaction centre, not a complex set of machinery. Warburg had indeed discovered the key enzyme involved in the transfer of electrons to oxygen – he won the Nobel prize for this discovery in 1931 – but was wrong about the context. The notion that respiration actually works by pumping protons across a membrane to generate a strong electrical potential seems never to have crossed his mind.

The same attitude coloured Warburg's view of photosynthesis, forming the basis of a long-running dispute with Robert Emerson about the number of photons of light needed to produce one molecule of oxygen. Emerson did his PhD with Warburg before moving to Illinois and assembling the 'Mid-West gang' of researchers interested in the mechanics of photosynthesis. Warburg originally claimed that only four photons were needed to release a molecule of oxygen, and later, in pursuit of some sort of thermodynamic perfection, he proclaimed a mere two. Emerson had been unwilling to challenge the master, but his own measurements indicated that as many as eight to twelve photons were needed. As usual, Warburg was not interested in wider discoveries and simply ignored evidence for the Z scheme (see Chapter 4). His position was again grounded in a vision of the perfection of physics, captured in this passage: 'The reaction by which nature transforms the energy of sunlight into chemical energy, and upon which the existence of the organic world is based, is not so imperfect that the greater part of the applied light energy is lost; on the contrary, the reaction is like the world itself, nearly perfect.' To my mind, this remark is in the same vein as Einstein's reaction to the quirkiness of quantum mechanics: 'God does not play dice.' The idea that life is nearly perfect is alien to evolutionary biologists, who see messy trade-offs everywhere. Yet Warburg saw nature through a prism that filtered out the tortuous complexity of life. Denouncing Emerson as wrong, Warburg's own measurements nonetheless eventually yielded an estimate of twelve photons too, yet he never admitted to any error. The problem, as in his prolonged dispute with Keilin, was that Warburg's vision of nature was coupled to a sense of his own infallibility and disdain for the ability of others. Certainly, he was a genius, but these failings diminish his stature as a scientist and human being.

We need to interpret Warburg's forceful pronouncements on cancer in this light. His first experiments were indeed striking, and set the scene for decades. In the early 1920s, Warburg showed that cancer cells produced *seventy times* as much lactic acid as 'normal' cells, even in the presence of oxygen. That is a big number, almost a physics number: two orders of magnitude! What does it mean? Lactic acid is yet another carboxylic acid, which readily loses a proton to form lactate. It is formed from pyruvate, as the waste product of fermenting glucose in animals; yeasts use exactly the same pathway, differing only in the final steps, to make the more welcome product ethanol instead. Here's what happens in animals, with pyruvate on the left picking up '2H' from NADH to form lactate on the right:

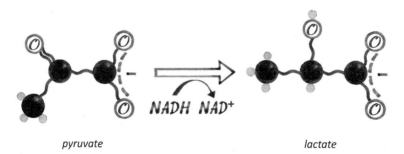

pyruvate   lactate

Most people will be familiar with lactic acid as the molecule associated with fatigue and muscle cramps in marathon runners, or lesser mortals over shorter distances. You can see that the middle 'mad-eye' oxygen has been replaced with an OH group – an alcohol. It is far less reactive but still somewhat intoxicating. The accumulation of lactate normally betrays a shortage of oxygen: an inability to make enough ATP through the Krebs cycle and respiration. Like respiration, fermentation also produces ATP, albeit barely a tenth as much. The breakdown of glucose to pyruvate ('glycolysis') normally produces two molecules of ATP, plus two molecules of NADH (which is to say the '2H', usually oxidised in respiration). The problem, when oxygen is unavailable, is that too much NAD^+ picks up 2H to become NADH:

$$NAD^+ + 2H \longrightarrow NADH + H^+$$

Why is this a problem? Because glycolysis cannot generate more ATP unless the NADH can be oxidised back to NAD^+, which can then pick up more 2H from the breakdown of glucose. That oxidation would normally be accomplished by respiration, but in the absence of oxygen it has to be achieved instead by converting pyruvate to lactate. Look again at my portraits opposite. The *raison d'être* for this reaction is to oxidise NADH, permitting glycolysis to continue, thereby generating a modicum of ATP. Lactate is a waste-product, washed out from cells, to be reused somewhere else in the body in better times. The bottom line is that fermentation enables a little ATP synthesis in the absence of oxygen.

Warburg was instrumental in piecing together the key reactions of glycolysis, yet cancer was already becoming his obsession. Why on earth would cancer cells switch to a profligate strategy of generating a tenth as much ATP as normal, while throwing away a three-carbon molecule that could be used for other purposes (such as biosynthesis) instead, even when there is plenty of oxygen available? Warburg was clear: the machinery of respiration had to be damaged. Cancer cells lost the ability to respire, reverting to what Warburg perceived as a lower form of life, lacking the structure and differentiation of our own higher cells. He described fermentation as 'the energy-supplying reaction of the lower organisms', and asserted that not even yeast, 'one of the lowest forms of life' could live by fermentation alone without 'degenerating to bizarre forms'. One can almost hear the language of the Third Reich. Warburg reduced the complexity of cancer to a clear vision. 'Even for cancer', he wrote, 'there is only one prime cause … the replacement of the respiration of oxygen in normal body cells by a fermentation of sugar.'

This view of the degeneracy of cancer, and the promise of a cure, probably saved Warburg's life during the Hitler years. Despite seeing himself as a Protestant German patriot, the Nazis considered him to be 50 per cent Jewish, and he was for a brief period relieved of his position. But he was swiftly reinstated, under Göring's declaration 'I will decide who is a Jew.' The reasons are shrouded in mystery, but Hitler apparently had a deep-seated fear of cancer. His mother, said to be the only person he ever truly loved, died from breast cancer when Adolf was eighteen. Her Jewish doctor, Eduard Bloch, wrote that 'outwardly, Hitler's love for his mother was his most striking feature. I have never witnessed a closer attachment' and 'in all my career, I never saw anyone so prostrate with grief as Adolf Hitler'. For some years, Hitler wrote postcards to Dr Bloch, and in 1937 granted Bloch and his family safe passage from Austria; they eventually arrived in the Bronx, New York.

Hitler's enquiries after Bloch might have been prompted by his own brush with cancer during the 1930s, when he had a growth removed from his left vocal cord in 1935 (and another in 1944).

This context no doubt contributed to the Nazis' tolerance of Warburg, who was never shy of promising salvation. In 1940 he expressed his hope for a third Nobel nomination, when he predicted the problem of cancer 'will be resolved within two years'. During the Nazi period, Warburg kept his head down; he was rarely critical of the regime, though he did try to help Krebs and others. But this apparent collusion did not endear him to other scientists. Krebs wrote that 'Warburg's willingness to let his Jewish blood be diluted in this way, and thus to make a pact with the Nazis, incensed colleagues outside Germany.' Krebs wrote a fond biography of his mentor, lauding his scientific accomplishments, but was equivocal on Warburg as a man: 'As Goethe said of Faust, he was one of those "who never ceased to strive". And thus, like Faust, he was a good man.' Faint praise; but perhaps that is the best any of us can hope for.

Rewiring metabolism

Warburg's position on cancer owed more to his philosophy than it did to his experiments, just as it did with photosynthesis. The basic problem with the Warburg effect – the propensity of cancers to ferment glucose in the presence of oxygen, or in the conventional shorthand, 'aerobic glycolysis' – is that it is far from universal. Many cancers don't depend on aerobic glycolysis at all, as Warburg's own measurements had shown, in agreement with those of his detractors. Normal tissues are also capable of aerobic glycolysis, to a similar degree; and stem cells (which give rise to new cells in tissues) typically depend on ATP from aerobic glycolysis for their energy needs. Some cancer cells even take up lactate from outside and burn it in their mitochondria, in what's called the reverse Warburg effect. So, the Warburg effect is certainly not a universal key to cancer. In which case, why has the number of papers citing it risen so swiftly over the last decade? Because Warburg was not entirely wrong either. He was onto something. Just not quite the right thing.

Warburg's tendency to assert truth in the face of conflicting data, while at the same time belittling the thinking of his opponents, climaxed in his famous *Science* paper in 1956. It's worth spending a moment on this because it shows how truth can lose its way through alienation in science, just as in life. Warburg hardly even deigned to throw down the gauntlet: 'The era in which the fermentation of cancer cells or its importance could be disputed is over, and no-one today can doubt that we understand the origin of cancer cells … if we know how the damaged respiration and the excessive fermentation of cancer cells originate.' Researchers of the calibre of Britton Chance and Sidney Weinhouse were quick to point out that these statements were not even backed by Warburg's own data. His self-righteous tone would make a saint bristle. Yet to dismiss Warburg as a monomaniac, which is what came to pass, overlooks the fact that his vision was coherent, even if it was partly misconstrued.

Warburg continued in similar vein: 'If the explanation of a vital process is its reduction to physics and chemistry, there is today no other explanation for the origin of cancer cells, either special or general.' This oblique reference to relativity seems intended to call upon physics as the only true science, and Warburg as the only physicist in the room. I hear the angry shuffling of feet towards the door. 'From this point of view, *mutation* and *carcinogenic agent* are not alternatives, but empty words, unless metabolically specified. Even more harmful in the struggle against cancer can be the continual discovery of miscel-laneous cancer agents and cancer viruses, which, by obscuring the underlying phenomena, may hinder necessary preventive measures and thereby become responsible for cancer cases.' At an emotional level, he may as well have said: if you don't agree with me, you are responsible for killing people. Yet if we disso-ciate this statement from its emotive context, there is some truth here. This is why there has been a renaissance in the Warburg effect. Mutations, carcinogens and cancer viruses do indeed all converge on the metabolic machinery. Of course, we wouldn't know that if thousands of researchers had not dedicated their lives to studying the very details that Warburg dismissed as obscuring the underlying phenomena. But now we know that many oncogenes, tumour suppressor genes and cancer viruses do indeed all induce some form of metabolic rewiring.

Warburg was arguably right that cancer is fundamentally a metabolic rather than a genetic disease, but his focus on cell respiration missed the main point. For growth is the key, and growth needs more than ATP. Respiration undoubtedly declines as we age, but the underlying problem is not a shortfall in ATP, but rather a shift in metabolic flux and gene activity towards a state favouring growth. Cell replication requires the doubling of all of a cell's contents, which means new sugars and new amino acids to make nucleotides for RNA and DNA, new amino acids for proteins, and new fatty acids for membranes, as well as suf-ficient antioxidant defences to escape any checkpoint controls that would otherwise trigger cell-death programmes (apopto-sis). Some of the substrates needed for growth are delivered in the blood supply, but most of them must be crafted from their precursors in the crucible of the cell itself. And that requires the rewiring of metabolism to produce everything that's needed for cell replication.

Let me give a single revealing example, from the pioneering cancer biologist Craig Thompson, who has done as much as anyone over the last decades to reinterpret the Warburg effect in terms of metabolic reprogramming. Thompson gives the example of the sixteen-carbon fatty acid palmitate, which is a major constituent of cell membranes. To produce a single molecule of palmitate requires seven molecules of ATP, but more critically in this context, it also needs sixteen carbons from eight molecules of acetyl CoA, plus twenty-eight electrons, which come from fourteen molecules of NADPH. I'll come back to NADPH in a while, but for now let's just note that this is *not* the same thing as NADH. Biochemistry can be irritatingly fiddly in this way, but you really do need to mind your Ps (and Qs). The 'P' in NADPH actually designates a phosphate, which allows it to interact with a different set of enzymes; but NADPH has more power to drive reactions, so you could think of the P as standing for the 'power' form of hydrogen.[3] As a rule of thumb, both NADH and NADPH transfer 2H, but NADH passes them on into the furnace of respiration, to generate ATP, whereas the power form, NADPH, drives most biosynthesis – making new molecules. That means NADPH is essential for growth and replication. It's also needed for producing a small peptide called glutathione, which maintains the antioxidant defences

---

3 The reason NADPH has more power than NADH has nothing to do with the structure, which is functionally identical, but relates to how far each one has been pushed away from its expected chemical equilibrium. If oxygen is present, both NADH and NADPH should be entirely oxidised to NAD^+ and $NADP^+$, respectively. In cells, however, NADH accounts for 20–30 per cent of the total inside mitochondria, and just 1 in 1,000 in the cytosol (see footnote 5 in Chapter 4). In contrast, NADPH makes up almost the entire $NADP^+$ pool: about 99.5 per cent is in the form of NADPH, with merely 0.5 per cent as $NADP^+$. This means that NADPH is much further from equilibrium (like an overfilled balloon), and so is capable of doing much more work.

of cells. We'll come to this in the next chapter; let's just note here that cells that can't make enough NADPH are more vulnerable to oxidative stress, and more likely to kill themselves by programmed cell death. Cancer cells escape this fate by making more NADPH. All in all, making NADPH is at least as important as making ATP for growth (and so cancer), and cells must wire their metabolism to get the balance between ATP, NADPH and carbon skeletons right.

Here's the problem. Let's start with glucose, which I'm sure you'll remember is a six-carbon sugar. A single molecule of glucose could produce between thirty and thirty-six molecules of ATP through aerobic respiration, but only two molecules of NADPH (by way of a pathway known as the pentose phosphate shunt); or it could provide six carbons towards the sixteen needed for palmitate. This means that one glucose molecule provides five times the ATP needed to make a single palmitate, whereas seven glucose molecules would be needed to generate the NADPH required, what Thompson calls a thirty-five-fold asymmetry. Three molecules of glucose are needed to provide the sixteen carbons for palmitate, which offsets this asymmetry a little, as only four of seven glucose molecules could be completely oxidised in respiration. But even after taking that into consideration, the seven molecules of glucose needed to make one molecule of palmitate would still provide nearly 200 ATPs, after siphoning off the carbons and NADPH needed for synthesising palmitate. Recall that just seven of these ATPs are actually needed, so respiring all the glucose not needed for biosynthesis would make twenty-eight times more ATP than needed.

Great, you might think, what's not to like? But in fact, there's a serious cost. Having excess ATP sloshing around actually *switches off* the glycolytic breakdown of glucose to form the NADPH and acetyl CoA needed to make palmitate at all. Likewise, flux through the Krebs cycle has to slow down if ATP is not being consumed fast enough. That's because the speed of respiration, and so the Krebs cycle, depends on the flow of protons through the ATP synthase; this flow jams up when ATP accumulates. Slow flux through the Krebs cycle restricts the availability of biosynthetic precursors and retards growth. Remember our assembly line analogy – if the product is not removed, it will pile up in the factory, and the assembly line has to be closed down until the product can be safely sold on. Likewise, you don't want a pile-up of ATP. If you were a cancer cell, the last thing you'd want to do is burn glucose through cell respiration, as it will produce a pile-up of ATP. The requirements for biosynthesis and ATP synthesis are just not closely aligned.

This picture is a little exaggerated, because fatty-acid synthesis is especially demanding of NADPH, but the synthesis of amino acids and nucleotides also consumes more NADPH and carbon equivalents than ATP. The synthesis of RNA, DNA and especially proteins has higher ATP demands (joining monomers together in chains costs ATP but not carbon or NADPH), although in this case the demand depends primarily on how many proteins are actually made. The workhorses of tissues are 'differentiated' cells, so called because they have plenty of recognisable cellular structures under the microscope. These structures are made largely of proteins, and protein synthesis has the highest ATP costs of all. Operating all this machinery makes eye-watering ATP demands too. For these reasons, once cells become differentiated to carry out specific tasks in tissues, they typically switch over to aerobic respiration to provide the ATP necessary.[4] But cancer cells cut back on these costs – they

---

4 Conversely, stem cells, which give rise to most new cells in tissues, are typically

lose their differentiation, and now spend less on making and operating proteins, while capturing more of the carbon and NADPH needed for growth. To grow, cancers need to reprogram their metabolism away from generating ATP through the Krebs cycle and respiration, commonly switching instead to aerobic glycolysis: the Warburg effect. To suppose that the Warburg effect reflects a primary degeneration of respiration misses the point. It is not about ATP – it's about growth. Cancer cells *must not* make too much ATP, as it *slows down* their growth. Cancer cells switch over to aerobic glycolysis precisely because it makes *less* ATP, favouring faster growth.

If that were all there was to it, I could stop here. We would have solved cancer long ago. Why would I even write about it in a book on the Krebs cycle, when we've seen that metabolic flux diverts towards glycolysis in cancer cells? I hope you'll guess the answer by now. From the very origins of life, the Krebs cycle was central to biosynthesis. Cells always had to balance energy against growth, a balance that was from the beginning grounded in the yin and yang of the Krebs cycle. It should have come as no surprise that mutations in Krebs-cycle enzymes could drive cancer, but still it did. And over the last few years, these mutations have taught us a lot about what goes wrong when a cell turns bad. That doesn't mean these mutations drive all forms of cancer. Rather, they provide a route map of central metabolism, which also begins to show what goes wrong as we age.

---

not differentiated, meaning there's little to see inside, and they tend to rely on aerobic glycolysis (fermentation) for energy, rather than respiration. In that respect, they are similar to cancer cells. As new cells differentiate, they typically switch over to respiration for their energy. The enzyme that allows them to do this is pyruvate dehydrogenase; we'll return to this vital regulatory enzyme in Chapter 6.

An ancient switch

The first mutations in genes encoding Krebs-cycle enzymes to be linked with cancer were found in succinate dehydrogenase and fumarate hydratase. Both these mutations block Krebs-cycle flow in the vicinity of the C4 intermediate succinate. I've never forgotten hearing about this for the first time from Christian Frezza in Cambridge, before I gave a talk in his department. I have no idea what I talked about that day – but the excitement of his perspective on the Krebs cycle in cancer has never left me. Since then, Frezza and a few of his colleagues around the world have reconceptualised the role of the Krebs cycle and metabolic flux in cancer.

Recall that at the end of Chapter 4 we touched on the origins of animals in stagnant muds beneath the late Proterozoic oceans, right before the Cambrian explosion. These early worms had to deal with low oxygen and excess hydrogen sulfide in their environment. Think what happens to the Krebs cycle under those conditions. In the absence of oxygen, NADH isn't easily oxidised, so it builds up. Just as in glycolysis, its accumulation obstructs normal flux. But we also saw that the Krebs cycle frequently operates as a two-pronged pathway, in which an oxidative branch that generates NADH is balanced by a reductive branch, which consumes NADH (see the schematic on page 185). The two forks join in the vicinity of succinate. And that's no accident.

In many animals that need to deal with low oxygen levels, a slight excess of NADH can allow for a little ATP synthesis. This partially explains, for example, how turtles can survive underwater for hours without breathing. When oxygen is not available, fumarate is used as an alternative electron acceptor, meaning electrons flow to fumarate rather than oxygen. That works because the enzyme fumarate reductase (which is similar in structure to our own succinate dehydrogenase) is embedded in the mitochondrial inner membrane – it forms complex II of the respiratory chain, the only enzyme in the Krebs cycle that physically sits in the membrane. Electrons can therefore flow from NADH in complex I to fumarate in complex II, giving succinate as the waste product. This electron flow allows the pumping of four protons at complex I, driving a little ATP synthesis in animals that possess the fumarate reductase enzyme. Overall, it works like this (where CI depicts complex I, CII is complex II and the right-hand complex is the nanomotor ATP synthase):

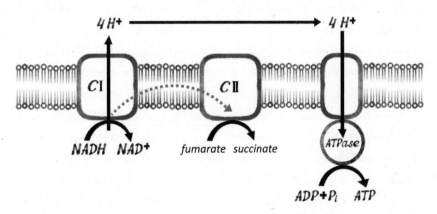

The dotted line shows the transfer of two electrons from NADH to fumarate. The waste product, succinate, accumulates, and can readily escape from the mitochondria. Now, here's the rub, which makes this ancient mechanism still relevant to cancer. Succinate is a powerful signal. That's because it is the only Krebs-cycle intermediate that can *only* be made in mitochondria (other intermediates can be formed elsewhere in the cell), precisely because the enzyme sits in the inner mitochondrial membrane, as complex II. This location means that succinate gives unique feedback on the overall status of mitochondrial function – on the ability of mitochondria to provide enough ATP to supply the needs of the cell. In short, if succinate is accumulating, then something is going wrong with respiration that needs fixing *right now!* Succinate is therefore far more than a waste product – it is an active signal that switches genes on or off.

Once in the cytosol, succinate binds to a class of proteins called prolyl hydroxylases, which blocks their normal action. The most interesting normal action of these prolyl hydroxylases targets another protein, 'hypoxia-inducible factor', or $HIF_{1\alpha}$, causing it to be degraded. Here's a fact to conjure with. $HIF_{1\alpha}$ is formed constitutively, which is to say that the gene encoding it is read off continuously to form the protein, but as soon as it's formed the protein is targeted and degraded by prolyl hydroxylases, *with a half-life of just five minutes*. Five minutes! Protein synthesis is an expensive and time-consuming process. Imagine you make a spanking new protein. What should you do with it? Don't think about using it – destroy it immediately! Then repeat, endlessly. Einstein once quipped that the definition of insanity is doing the same thing over and over again and expecting different results. But that's exactly what virtually all cells do, virtually all the time. And so long as oxygen is available, they get the same result every time. But if oxygen levels fall, succinate builds up and blocks the breakdown of $HIF_{1\alpha}$. And now $HIF_{1\alpha}$ finds its way to the nucleus, where it binds to DNA and switches on hundreds of genes. All that work and trouble serves just one purpose – so you can be damn sure it's a critical one. It buys the cell just enough time to fix the problem before it suffocates to death. $HIF_{1\alpha}$ is made continuously just to ensure that fully functional (spanking new) copies of the protein can make it to the nucleus in time, in the rare event of an oxygen emergency.

So how does $HIF_{1\alpha}$ making it to the nucleus in time save the cell from certain death? Look at the genes that it targets. They include genes for glycolysis, which we've seen promote ATP synthesis in the absence of oxygen. But the genes switched on by $HIF_{1\alpha}$ also encode proteins that drive growth and inflammation. When I talked about low oxygen, you might have imagined putting a plastic bag over your head, or drowning desperately, but those catastrophes are far too sudden for changes in gene activity to fix. That's not why this machinery is necessary. Think about the conditions that are normally associated with low oxygen levels. For most animals, the most predictable life-threatening form of hypoxia is infection. Proliferating bacteria and immune cells consume oxygen faster than it can be provided, leading to swelling, damage and the partial occlusion of capillaries. So the genes switched on by $HIF_{1\alpha}$ aren't limited to those that deal with low oxygen: they orchestrate the whole inflammatory response too. Inflammatory signals promote the growth of new blood vessels, the proliferation of immune cells, and the resistance of bystander cells to more cell death. In short, $HIF_{1\alpha}$ helps coordinate a response to low oxygen levels that stimulates cell growth, upregulating glycolysis and balancing the cell's needs for ATP, NADPH and carbon. If these pro-growth, pro-survival, pro-inflammatory conditions persisted for too long, you can see they might promote the uncontrolled growth of cancer cells.

How could such conditions persist? Let's return to succinate itself. Our own cells can't make ATP from the transfer of electrons to form succinate, as the first animals did, because our ancestors lost the gene for fumarate reductase. Presumably they no longer needed to deal with high levels of hydrogen sulfide in their environment. We abandoned those stagnant muds and ended up on land, breathing air that fairly crackles with oxygen. Yet our equivalent enzyme, succinate dehydro-genase, can also switch direction and transfer electrons from fumarate to succinate, so we can accumulate succinate in the same way, even if we can't generate ATP on the side. In fact, the interconversion of succinate to fumarate is exquisitely close to equilibrium, meaning that it flips from one way to the other quite readily, depending on the concentration of each reactant, the electrical charge on the membrane and the amount of ATP relative to ADP. In short, succinate gives a delicately balanced real-time feedback on the status of the respiratory system as a whole. Even slight impairments in Krebs-cycle flux give a read-out of succinate levels, which are counteracted through changes in gene activity. The list of medical conditions where this mechanism turns out to be relevant is sobering. I'll come back to a couple of these in the next chapter but just think for a moment about the common disorders where oxygen delivery to tissues becomes compromised. Stroke. Heart attacks. Demen-tia. Arthritis. Organ transplants. Succinate accumulation is central to them all.[5]

---

5 My own PhD, back in the 1990s, addressed problems with oxygen delivery in organ transplantation. I found that when a kidney was transplanted after a few days of storage, the reintroduction of oxygen caused the mitochondria to go haywire. NADH was oxidised very slowly, and its electrons didn't seem to make it down the respiratory chain to cytochrome oxidase and oxygen. I never figured out what was going wrong. Twenty years later, Mike Murphy, Christian Frezza and colleagues in Cambridge elucidated the mechanism in just beauti-ful detail. In organs stored for transplantation, as in myocardial infarctions (heart attacks), succinate builds up while oxygen is restricted. Then, upon the return of oxygen (reperfusion) the succinate is rapidly oxidised, its electrons flooding into complex II. These excess electrons in effect swamp the system and hinder the oxidation of NADH at complex I, producing a burst of reac-tive oxygen species (ROS) that can damage the machinery of respiration – the self-destruction of mitochondria that I struggled to explain. We'll think more about ROS in Chapter 6.

You'll appreciate that a mutation in the enzymes succinate dehydrogenase or fumarate hydratase will cause an accumulation of succinate in cells that are otherwise perfectly healthy. Plainly, damming metabolic flow will form a reservoir behind the dam:

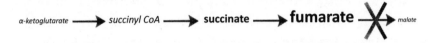

You might call this fake news. The mutation makes the cell think it needs to survive in a low-oxygen environment, even when there's plenty of oxygen around. I should say that the roles of fumarate and succinate overlap in terms of their signalling effects. Both stabilise $HIF_{1\alpha}$ and related proteins, giving a suite of 'epigenetic' effects (meaning they can silence some genes and amplify others). Let's not get lost in detail, but just note that these mutations promote a switch to glycolysis even in the presence of oxygen – the Warburg effect – which drives cell growth. The requisite machinery was there all along, serving the same purpose that it did in early animals, but switched off until needed. Mutations in Krebs-cycle enzymes set off this machinery in the wrong context, with no obvious way of closing it down again (as the switch is disconnected from oxygen), increasing the risk of cancer.

Increasing the risk … but nothing is set in stone. More is needed. Mutations in succinate dehydrogenase promote the switch to aerobic glycolysis, but that's only half the story. Let's keep following the money. Cell replication is not only about ATP, NADPH and carbon. There's also an imperious need for nitrogen to make new amino acids for proteins, and nucleotides for RNA and DNA. While there's some nitrogen turnover in cells, for the most part, nitrogen is delivered around the body in the form of the amino acid glutamine. And that, too, is linked to the Krebs cycle.

Another reversal

Most cancers need no more than glucose and glutamine to grow. These substrates serve all of their needs. And of the two of them … give the cancers glutamine. I have a vivid memory of discussing glutamine and the Krebs cycle with Salvador Moncada in his directorial office at the Wolfson Institute a decade ago. Moncada had few equals in bestriding pharmaceutical and academic research. He arrived at the question of cancer after decades of working on the vascular system, culminating in the simultaneous discovery, by two different labs, that the reactive gas nitric oxide (NO) dilates blood vessels, a process that was to become the basis of a blockbuster drug for erectile dysfunction, Viagra. Like carbon monoxide (used by Warburg to identify the haem pigment in his 'respiratory ferment'), nitric oxide binds to cytochrome oxidase, blocking respiration and modulating programmed cell death. These findings led Moncada to explore the role of mitochondria in controlling the cell cycle.

Cancer cells growing in culture can be checked in their mad career through the cell cycle by depriving them of glucose or glutamine. Around the time we were talking, Moncada had just shown that depriving cells of glucose did not pause their growth for long, but depriving them of glutamine put paid to their prospects permanently. And now here he was, chalk in hand, eyes sparkling, drawing arrows spiralling out from the Krebs cycle into the cytosol. Flux in reverse to citrate. Export, then break down to acetyl CoA and oxaloacetate. Thence pyruvate and lactate. Not from glucose at all, but glutamine. And on. It was dizzying, a prelude to the coming storm. I regret the storm was not of Moncada's making. He left UCL soon afterwards, to direct a new Institute for Cancer Sciences at the University of Manchester; but whenever I think of glutamine,

I can't help but see Moncada pacing urgently before his black-board. Molecules acquire personality this way.

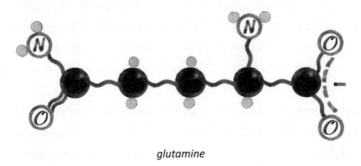

glutamine

The first wave of the storm came from Navdeep Chandel, Ralph DeBerardinis and colleagues in 2012. They reported in *Nature* that cells with defective mitochondria would habitually drive part of their Krebs cycle backwards, to make citrate (C6) by reacting α-ketoglutarate (C5) with CO_2 – a rare case of carbon fixation in animals, equivalent to photosynthesis without the sun. As I noted in Chapter 2, this reaction was discovered long ago, by the great Spanish biochemist Severo Ochoa, in New York in 1946, and had become well known in bacteria as part of the reverse Krebs cycle. But what was unusual here was the sheer scale. This reverse flux is the *dominant* pattern in cancer cells with mutations in fumarate hydratase, or more widely in cells with damaged respiration, caused by mutations or drugs that impair electron transfer. Make a mental note of that: damaged respiration ... roughly as predicted by Warburg. That will be important, but we need to understand how the Krebs cycle can go into reverse first.

Once again, cancer cells take advantage of existing machinery, and in this case subvert a marvellous mechanism, a mechanism that showcases both the glory of natural selection and the cleverness of cancer. For good measure, it also relates to NADPH, the 'power' form of hydrogen, which I hope you'll recall drives biosynthesis and antioxidant defences.

The enzyme that catalyses the interconversion of isocitrate into α-ketoglutarate is known as isocitrate dehydrogenase. This comes in two main forms, each of them encoded by a different gene. The standard form does what it says on the tin – it removes 2H, in the form of NADH. The NADH is then fed into the furnace of respiration at complex I; this is the conventional Krebs cycle (see page 286). More intriguingly, the second form of the enzyme does the opposite. But going in this reverse (reductive) direction is more difficult, and so requires the powerful NADPH to cobble its 2H onto α-ketoglutarate. Even then, it can only power this reverse step when NADPH is abundant. If NADPH levels dwindle (by being oxidised back to NADP⁺), then the enzyme flips back to the forward (oxidative) direction. That generates NADPH instead of NADH, but otherwise it works in the same way as the standard form of isocitrate dehydrogenase.

What's the point of that, you may be wondering? Well, it works as both a release valve and a spring-loading mechanism, allowing us to leap into explosive action. Imagine that you're resting (OK, you're sitting around reading this book), so demand for ATP is low. What's going on inside your cells right now? Everything in your mitochondria is linked in series. If demand for ATP is low, then ATP is not being consumed, which means that the ATP synthase stops spinning. So protons no longer flow through the rotary motor of the ATP synthase and remain outside. That supercharges the electrical membrane potential. Respiration can't pump protons against a supercharged electrical potential, so electron transfer to oxygen slows right down. And that means NADH can't be oxidised, so the Krebs cycle has to slow down too. Everything grinds to a halt. Reading, eh!

This combination of high levels of NADH with super-charged membrane potential raises the threat that electrons will escape from the respiratory chain and react directly with oxygen, forming reactive oxygen species (ROS). These can damage lipids, proteins or DNA. I can hear the threatening buzz of electricity. If this were a Bond movie, klaxons would be sounding and an urgent voice would be repeating 'Danger! System overload! Evacuate building!' But don't worry, it's not going to blow. Evolution outsmarts the average Bond villain.

Under these conditions, the supercharged membrane potential powers the conversion of NADH into NADPH, through a 'transhydrogenase' enzyme sitting in the membrane (which takes its power, like the ATP synthase, from a flow of protons). The transhydrogenase transfers 2H from the excess NADH onto $NADP^+$. That sounds so anodyne that you might have missed just what a beautiful, simple solution this is. The transfer simultaneously lowers the membrane potential and NADH levels, while topping up NADPH, bolstering antioxidant defences. So it acts as a release valve, alleviating all three components of the risk (supercharged membrane potential, excess NADH and ROS). But notice that for this mechanism to continue, we need to regenerate the $NADP^+$, so it can keep on picking up 2H from NADH. That brings us back round to the second, NADPH-linked form of isocitrate dehydrogenase. Remember that high NADPH levels drive this enzyme backwards, meaning it converts α-ketoglutarate to isocitrate and onwards to citrate. Reverse flow through the Krebs cycle keeps the right balance between $NADP^+$ and NADPH. We'll come back to what happens to the citrate soon, but for now let's just appreciate that it can be exported out to the cytosol and used for making fats. Reading, eh!

First, let's consider what happens if there's now a renewed

demand for ATP, a sudden call to action. You decided to go for a run. As ATP is spent, the ATP synthase springs into action, with protons flooding though its rotary motor. As protons flow back inside, membrane potential falls. Now NADH can transfer electrons into the respiratory chain again, so NADH levels fall. The shift in the balance between NADPH and NADH flips the direction of transhydrogenase enzyme. This now transfers 2H from NADPH onto NAD^+, which raises NADH levels. Rather than drawing on the proton-motive force, the flipped transhydrogenase pumps protons across the membrane, recharging the electrical potential. For the *coup de grâce*, as NADPH levels fall, the NADPH-linked isocitrate dehydrogenase flips direction too, so both forms now work in concert to convert isocitrate into α-ketoglutarate, powering the Krebs cycle in the forwards direction. This is the spring-loading mechanism. In short, when there is a sudden need for explosive action, the Krebs cycle is powered by multiple engines working together, ramping up its power.

But hold on! This magnificent spring-loaded power driver has a weakness that cancer cells exploit – our Bond villain has a final dastardly trick up his sleeve. Think about the problem faced by cancer cells with a mutation in Krebs-cycle enzymes such as fumarate hydratase. The mutation blocks the Krebs cycle, so it becomes difficult to replenish the six-carbon citrate through the usual 'forwards' direction. I intimated that citrate is needed for making fats, and is exported from the mitochondria to the cytosol. There it is broken down into the C2 acetyl CoA, plus the C4 oxaloacetate.[6] The acetyl CoA is used to make fatty

---

6 The enzyme that breaks down citrate in the cytosol (in animals) is an ATP citrate lyase – which is related to the enzyme found in bacteria that drives the classic reverse Krebs cycle. There's nothing new under the sun. Curiously, plants don't do this – they generate the acetyl CoA needed for fatty acid synthesis

acids, which are needed for membranes, a vital component of cells, hence absolutely necessary for cancer cells to grow. They *need* that citrate.

You can probably see where I'm going with this. When forwards flux through the Krebs cycle is impaired, cancer cells can make citrate by converting α-ketoglutarate into isocitrate, then citrate, through reverse flux. You may be wondering where the α-ketoglutarate comes from, and the answer brings us right back round to where we started this section. Glutamine – the amino acid essential for cancer cells to grow. When the nitrogen is transferred from this five-carbon amino acid to make other amino acids and nucleotides, the carbon skeleton that remains is the C5 α-ketoglutarate. High levels of glutamine push the Krebs cycle into reverse by increasing the concentration of α-ketoglutarate. Mutations in the NADPH-linked form of the enzyme isocitrate dehydrogenase reinforce this shift in cancer cells, but these mutations are selected because they lock in a flux that is already beneficial for cancer cells. Mutations keep the enzyme jammed in reverse mode, so it keeps on making citrate, supporting cancer cell growth. Overall, the flux from glutamine looks something like the diagram on the next page:

---

in chloroplasts instead, from pyruvate directly, as in the normal Krebs cycle. Picking up endosymbionts such as cyanobacteria (which evolved into chloroplasts) gives the host cell an extra set of metabolic genes, with greater flexibility for compartmentalisation. Some algal cells with chloroplasts lost the ability to photosynthesise altogether, but retained their chloroplast anyway (now called a plastid) which they use for other purposes. A famous example is the malarial parasite *Plasmodium falciparum*, which has a plastid derived from its former chloroplasts. This has retained some genes that derive from the original algal endosymbiont, making them an interesting target for antimalarial drugs (as we don't have genes derived from algae, limiting the possibility of side effects).

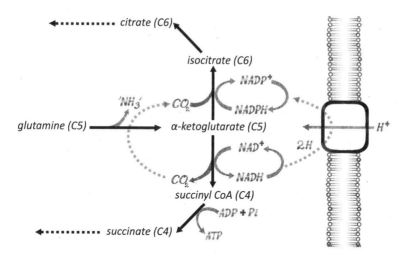

How glutamine sustains cancer growth: the 'NH₃' can be lost as ammonia or transferred to amino acids and used for nucleotide synthesis. The α-ketoglutarate can form both citrate (for biosynthesis) and succinate (which sustains the Warburg effect). The enzyme sitting in the membrane is the transhydrogenase, which regenerates NADPH from NADH, so long as there is a membrane potential.

We are left with a peculiar situation, in which Krebs-cycle flux is running backwards through large parts of the cycle; nothing like the canonical picture in textbooks. Far from being geared to ATP synthesis, the cycle is now exporting succinate and citrate from the mitochondria out to the cytosol. Together, these keep the levers for growth pushed hard down in the 'on' position. We've already seen how succinate promotes the Warburg effect, so now let's see what citrate does.

Turn left (or right) at Krebs junction

Once citrate is exported out to the cytosol, it is broken down into the C2 acetyl CoA, plus the C4 oxaloacetate. Acetyl CoA is used for fatty acid synthesis, but it also binds to proteins that control access to DNA, called histones. This enables many genes to be switched on: an 'epigenetic switch' known as acetylation. The genes switched on drive cell growth and proliferation – another signal that says *grow!* This epigenetic switch is reinforced by the context. The sirtuins are a group of enzymes that strip out the acetyl group, switching off the growth signal. But their activity depends on the availability of NADH. The sirtuins normally close down growth and sexual development when times are lean, and there's little NADH from food. Conversely, if NADH levels build up, which might happen if we eat too much and don't exercise, the sirtuins close down. They no longer strip out the acetyl groups, permitting the growth signal to persist. The problem in cancer is that reverse Krebs-cycle flux triggers the same signal, because cancer cells have lots of NADH from aerobic glycolysis, and lots of acetyl CoA from citrate. So the growth switch remains jammed on.

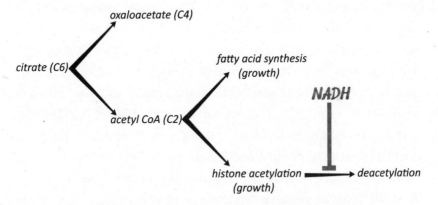

How export of citrate from the mitochondria can drive growth through an epigenetic switch. When NADH is plentiful, the sirtuins are suppressed. Because they would normally switch off the growth

signal by deacetylating histones, the combined effect
of acetyl CoA and NADH is to promote growth.

The other breakdown product from citrate, oxaloacetate, is equally important. It stands at a unique crossroads in metabolism, where almost every direction is good for cancer, which means the more you make, the better it gets. The choice is a bit overwhelming (see the figure opposite). You can pick up nitrogen to become an amino acid called aspartate, needed for nucleotide synthesis and to form urea (via the urea cycle that first brought Krebs to fame in 1932). Or you can be converted to the C4 malate, then stripped of CO_2 and 2H by the malic enzyme, to form the C3 pyruvate, plus (worth its weight in gold) the powerful NADPH, needed for biosynthesis. From pyruvate, you can form lactate and be washed out as waste. Much of the lactate excreted by cancer cells comes from glutamine in this way. Alternatively, from pyruvate, you can be broken down to acetyl CoA. And we've seen the havoc that can cause.

You're not out of choices yet. You can be stripped of CO_2, then strapped to a phosphate to form the C3 phosphoenol-pyruvate, the starting point for sugar synthesis via gluconeogenesis. This pathway is another ancient backbone of metabolism, more or less the opposite of glycolysis, which forms glucose. Hang on a minute … don't we want to burn glucose? No! Glucose is the starting point for making other sugars, most notably the ribose and deoxyribose needed for RNA and DNA, as well as more NADPH (via the pentose phosphate pathway mentioned earlier). And the path to glucose is replete with other useful intermediates, first among them the C3 sugar glyceraldehyde phosphate. That's the starting point for making glycerol phosphate, which binds to fatty acids to form the main phospholipid components of cell membranes. You can't grow without membranes.

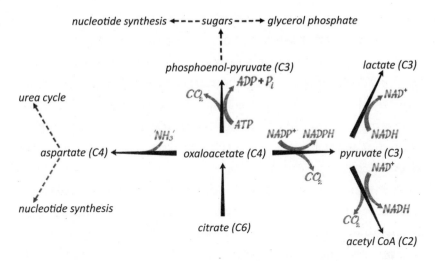

Krebs junction – oxaloacetate is at a crossroads in metabolism, with paths leading to nitrogen or sugar metabolism, or generating NADPH for biosynthesis and antioxidant defences. This is a key crossroads in energy metabolism; you see why Krebs was uncertain about the fate of oxaloacetate.

Glycerol phosphate has another equally important use. It can be used to generate electrical membrane potential in mitochondria, without involving the Krebs cycle. It works like this. Glycerol phosphate is stripped of 2H, which is fed into complex III of the respiratory chain from outside the mitochondria. These electrons are transferred onto oxygen in the normal way. Glycerol phosphate can be recycled by picking up 2H (from NADH in the cytosol) again and again. Overall, this unusual form of respiration oxidises NADH in the cytosol, passing its electrons onto oxygen, bolstering the power of aerobic glycolysis. Only six protons can be pumped this way (still enough to make a couple of ATPs) but there are two big advantages. First, making *less* ATP better matches the demands of growing cells, as we've seen, and the cycle can easily be switched off when it's not needed. Most tissues can get most of the energy they need most of the time through this mechanism, perhaps over many decades. And second, this peripheral cycle can drive the trans-hydrogenase enzyme to form NADPH inside the mitochondria, sustaining reverse flux through the Krebs cycle. Overall, your choices at what I'll call the Krebs junction look like the figure on page 225.

No wonder cancer cells are addicted to glutamine! Anything glucose can do, glutamine can do too, and it also brings nitrogen to the table. Glutamine even works as a self-fulfilling prophecy. I mentioned that one of the uses of oxaloacetate is to make aspartate for the urea cycle. In fact, the urea cycle is the only pathway here that is no use to cancer cells. Quite the opposite. It's long been puzzling that much of the nitrogen from gluta-mine is not bound into amino acids or nucleotides in cancer cells, nor is it safely incorporated into urea as a waste product, but rather it bubbles away as toxic ammonia. In a chilling sce-nario, it seems that the release of ammonia from glutamine in cancer cells actually promotes the breakdown of proteins in distant muscles. This works because glutamine is unusual in having two amino groups. To mop up the ammonia, muscle proteins break down and release their amino acids. The nitro-gen from these amino acids is transferred onto α-ketoglutarate to form glutamate, which then mops up the excess ammonia to form glutamine with its two amino groups. The glutamine is then recirculated in the blood – straight back to the cancer. (See the figure on the next page.)

Overall, raised levels of ammonia circulating in the blood are mopped up by the synthesis of new glutamine, through muscle breakdown, hiding behind the medical term cachexia. This is one of the deepest horrors of cancer, as the disease eats

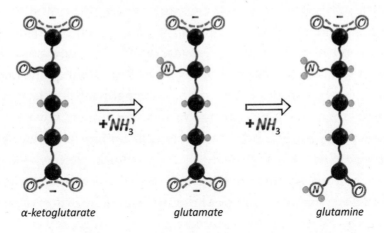

α-ketoglutarate   glutamate   glutamine

Formation of glutamine from α-ketoglutarate.
Note that the first 'NH$_3$' used to form glutamate
derives from amino acids, while the second, to form
glutamine, comes directly from ammonia.

away at our very fabric, strip mining the body with profligate disregard for greater good, to fuel its own selfish growth. From the cancer's point of view, always in the here and now, cachexia maintains a copious supply of free glutamine. It's as if a state provides free fossil fuels for the filthiest polluters. Cancer cells often acquire mutations in genes for the urea cycle. That way, excess nitrogen is splurged as ammonia rather than urea, promoting this vicious cycle while at once diverting more aspartate into the synthesis of nucleotides for RNA and DNA. If ever a cycle deserved to be called vicious, this is it.

There are many other ways in which cancer cells can evade textbook metabolic flux. There is no common pattern, although what I've described here is arguably as close to an 'archetypal' cancer metabolism as one could get. How cancer cells grow depends on what they're fed and whereabouts they are, even within the confines of a single tumour. For example, some cancer cells release lactate as a waste product, while others take it up. From lactate you can make pyruvate, then oxaloacetate and acetyl CoA. Everything I've just described is now possible, except that these cells must still get their nitrogen from somewhere (and are more likely to have normal Krebs-cycle flux).[7] In each case the rule is always 'follow the money'. Given any particular substrate, how do cancer cells find the right balance between NADPH, ATP, nitrogen and carbon skeletons to drive growth? Probably the only substrate that cancer cells may struggle to deal with are ketones. These are formed from acetyl CoA, either by breaking down stored fats, or through a 'ketogenic diet'. Because our physiology links ketone bodies with starvation, their presence tends to switch off growth signals. It's also

---

7 Books tend to take their own direction against the best-laid plans of the author. I wanted to write about the work of Lee Sweetlove in this chapter, but sadly it only made it into this footnote. Probably only a tenth of the material I wanted to write about actually made it into the book. Sweetlove's 2010 paper 'Not just a circle: Flux modes in the plant TCA cycle' opened my eyes to the sheer variety of flux patterns through the Krebs cycle, all linked to growth and with huge relevance to cancer – not despite, but because, it was about plants rather than animals. Plants don't need as much ATP from their mitochondria, as photosynthesis generates plenty of ATP in chloroplasts. But they still need mitochondria to drive growth: for the carbon skeletons in the Krebs cycle, as described in this chapter. In fact, plants often solve the problem differently to animals – they tend to speed up forwards flux through the cycle by 'uncoupling' the respiratory chain (using an alternative oxidase or uncoupling proteins), to produce heat rather than ATP. No doubt animal cells can do this too, but it may be that too much heat can kill cancer cells, or indeed human beings. In any case, plant scientists were routinely thinking about the Krebs cycle in terms of growth, rather than ATP synthesis, long before most cancer biologists began to think that way. We should never forget that ideas in science often come from unexpected quarters. Focusing funding on research into human cancer would miss profound insights deriving from plant scientists. In a similar way, remember that the gene editing technology of CRISPR was discovered by scientists studying bacterial immunity to viral infections. Who knew they had an immune system?

notoriously tricky to make sugars from fats (although it can be done via the C3 intermediary acetone). Yet even these metabolic hurdles are not insurmountable for cancer cells.

I've implied that cancer is primarily a metabolic disease, yet if anything goes for metabolism, then what actually turns a cell cancerous? You've probably appreciated by now that the underlying problem is an environment that continuously and erroneously shouts '*grow!*'. This toxic environment can be induced by mutations, infections, low oxygen levels ... or the decline in metabolism associated with ageing itself. Remember that ageing is the single greatest risk factor for cancer. That's the main reason why I've talked so much about reverse flux through the Krebs cycle. I'm not trying to explain every case of cancer, but why our risk increases with age. Because age hits us all in the end.

The rising tide

Why, then, are we more likely to get cancer as we get older? I said at the beginning of this chapter that the answer does not lie in the accumulation of mutations with age. We've known for decades that they do not accumulate fast enough, nor do they account for the known patterns of cancer. Of course mutations *can* cause cancer, especially in younger patients (who are more likely to have inherited a predisposing gene) or those who smoke, for example. Mutations certainly accumulate in cancers, too, and have given us detailed insights into the pathways involved, including the Krebs-cycle mutations we've discussed. But they do not tell us why an older organism is more likely to succumb to cancer than a younger one, whether rat, human or elephant. That's a problem with ageing itself.

How old are you? Can you still run down the street and back

as well as you did when you were sixteen? Age catches up even with Olympic athletes. This is not a book about ageing, though we'll touch on a few of the mechanisms in the next chapter. But for now, let's just think about one thing – energy levels. As we get older, our respiratory performance declines slowly. Imperceptibly but inexorably. The rate of respiration is depressed the most at complex I, the largest and most complex of the respiratory complexes. That matters for two reasons. First, complex I is the main source of reactive oxygen species (ROS) from mitochondria, and the rate at which these escape (ROS flux) tends to creep up with age. We'll see in the next chapter that the rise in ROS flux can be compensated somewhat through suppression of complex I. That brings us to the second problem. Complex I is the beginning of the longest route through the respiratory chain, and the only entry point for NADH deriving from the Krebs cycle. Electrons enter from NADH and pass on down to oxygen, powering the extrusion of ten protons along the way. We saw in Chapter 1 that's enough to make three (and a bit) ATPs. So the decline in complex I activity with age means that it's no longer so easy to oxidise NADH.

Think what happens when NADH begins to accumulate in the mitochondria. The Krebs cycle must slow down, and there will be a tendency for sections of it to run backwards: if the NADH is not oxidised fast enough, then it will tend to push the reactions that produce it in the opposite direction. What happens? The ancient reductive fork of the Krebs cycle does what it normally does at low oxygen levels, and flips into reverse ... oxaloacetate to malate, then fumarate and succinate. The succinate accumulates, which signals an oxygen deficiency. There isn't really an oxygen deficit – it's fake news – but any deficiency in respiration signals essentially the same thing: upregulate glycolysis. Ferment glucose. Do a Warburg.

Meanwhile, forward flux from citrate to α-ketoglutarate and succinyl CoA has to slow down too. Unable to get much further through the forward cycle, citrate dissipates to the cytosol, where it is broken down into acetyl CoA and oxaloacetate. Histones, the gatekeepers to DNA in the nucleus, become acetylated. The epigenetic switch is pressed down, promoting growth, cellular proliferation and inflammation. Local inflammation can damage the surrounding tissue, but if it persists it will sap your appetite and weaken muscles more widely. Protein fibres begin to break down, releasing amino acids that get packaged up as glutamine and shipped off around the body for repurposing. You begin to fix CO_2 and other parts of your Krebs cycle will flip backwards. Ironically, you will put on weight by absorbing CO_2 and turning it into fats. Life's really not fair.

I might be in this state now, but to the best of my knowledge, I don't have cancer yet. Who knows. I worry about it in part because my dear friend Ian Ackland-Snow died from cancer in 2019, at the age of fifty-six, a year or two older than me. We sat together in the garden of the Royal Marsden Hospital, where he had had the best treatment anyone could wish for. He was a doctor himself and worked in medical education, a man of restless and questioning intellect, exuberant energy and a smile to light a room, who taught me much about how to write and think. Full of zest. Always active. A man who would run up mountains for fun. He was acutely aware of his chances, and exactly what we know or don't know about the treatments he was receiving; but also, how many imponderables there are. How much uncertainty about elementary questions. We have a long way to go before we can answer all these questions and face a diagnosis of cancer without fear. In the meantime, I hope I can muster Ian's fortitude and good humour when my time comes.

But why Ian? Why not others, who may smoke for decades,

spurn their greens and yet live to ninety? I can do no better than statistics. Ageing itself raises our risk, by switching metabolism towards aerobic glycolysis, promoting cellular growth. But clearly this state is stable over decades. We are then at the mercy of our own lives – unfortunate genes, one cigarette too many, poor diet, bad sunburn, exhaust fumes, viral infection: sharp focus for the command '*grow!*', *set in a permissive metabolic context*. If my argument in this chapter is right, the best we can do is keep our mitochondria active. Keep exercising. Breathe deeply. Eat carefully. Don't fall back on fermentation but oxidise NADH in your mitochondria as far as possible. Keep your Krebs cycle moving forwards. Nothing is failsafe, but there's no doubt that regular aerobic exercise and a healthy diet will help to protect you against cancer. There's irony in my advice, for I am asking you to nurture your mitochondria. Don't let cell respiration run down, which is the underlying cause of cancer as we age. Not because our cells revert to some degenerate state, but because declining respiration perturbs the Krebs cycle. It may not have been for all the right reasons, but Warburg was right after all. 'Do not go gentle into that good night ... Rage, rage, against the dying of the light.'

THE FLUX CAPACITOR

'So how do you explain me, then?' The question is lobbed like a hand grenade from the audience, intended to devastate the 'expert' standing on the stage. It's a question I've received more than once, typically from an old man, a lifelong smoker who last took any form of exercise decades ago. I have no idea whether they listened to what I had to say, or turned up only to make the speaker squirm. For it's plain that science is less good at explaining the particular than it is at statistical generalisation. A simple non-answer might be 'good genes', but that's little better than saying 'fate smiled upon you'. Of course, the questioner knows that. The grenade explodes hubris.

In fact, there is a deep problem about the general and particular that applies to every one of us, which arguably makes life tolerable. We all age: it is a completely generalisable phenomenon. With age, we all succumb to ailments or diseases. Yet one person suffers from cancer, another from Alzheimer's disease, another from a heart attack or stroke, or rheumatoid arthritis. Others die peacefully in their sleep with nary a varicose vein. Who knows what fate holds in store for each of us. Certainly, we can influence our fate through careful diet, exercise, meditation or even cold baths, but there's still something of a lottery about how we each meet our end. I do think life would be intolerable if it were otherwise – if we knew the appointed hour and cause of our demise. In this sense at least, ignorance is bliss.

Yet there is a standard answer, rooted in our genes, which goes back to the great immunologist Peter Medawar. We age, said Medawar, because we outlive our allotted time. Not allotted by God, but the statistical laws of selection. At some point we will be hit by a bus, or eaten by a bear, or be sabotaged from within by a virus. Selection had better focus resources on the period when we are likely to have offspring; there's no point preparing for a time that will almost certainly never come. So our genes are honed by selection to optimise fitness in our youth, with the strength of selection tailing off as we get older. Variants of genes that cause harm, which would be purged by selection if their effects were felt earlier in life, are less likely to be eliminated with age. Obviously, a gene that causes harm at the age of 150 cannot be rooted out by selection because none of us survive to that age. According to Medawar, much the same applies to variants of genes that cause harm at the age of seventy or eighty, as few, historically, lived to that ripe old age. The same argument applies, albeit with lesser force, to variants of genes that harm us in our forties or fifties, causing conditions such as Huntington's disease. This textbook view sees ageing and the diseases of old age as little more than the unmasking of late-acting genes, whose effects do us in. The conception provides a neat explanation for why we all age, yet suffer from different diseases of old age: the strength of selection falls with age universally, but we contend with our own unfortunate hand of cards – variants of genes, or alleles – giving us each our own unique blend of personal misery.

All of this is true, and if there is a picture from evolutionary biology that underpins modern medicine, then this is it. Different alleles predispose us to diverse age-related diseases. They cause their effects primarily in specific organ systems, hence the tendency of one system or another to let us down. The effect of these alleles is offset or exacerbated by hundreds, nay thousands or even millions, of other genetic variants, often merely single-letter changes in DNA here or there, known as single-nucleotide polymorphisms or SNPs (pronounced 'snips'). On average, we have one SNP every thousand letters, meaning that there are four or five million letters that differ (in various combinations) across the human genome as a whole. Only a modest proportion of these are likely to influence the risk of a particular disease, but they can be exposed by genome-wide association studies ('GWAS' for short) that assign specific SNPs some level of risk in predisposing to a particular disease. I imagine the genome as a vast kaleidoscope, each of us with our own unique, shimmering pattern of genetic variants that shape our individuality and craft our demise.

But (there's always a but) there are some serious problems with this view too. Think about 'our allotted time'. In the most general terms, this is a measure of chronological time. Our lifespan is determined by our likelihood of surviving a certain number of years – a statistical risk determined by our size, lifestyle, vulnerability to predators or infections, and so on. This is written into our genes, meaning that if we do outlive our expected chronological time (protected by the wonders of society), then we will be afflicted by a maelstrom of implacable spectres that consume us from within. To reverse the afflictions of older age would be to counter the effects of hundreds, nay thousands or millions of SNPs, a task tantamount to Sisyphus rolling his boulder up the mountain: if not strictly impossible, at least formidable.

So it came as a shock in the 1990s to discover that a handful of genes could double or treble the lifespan of simple creatures such as nematode worms or flies. Even mice. This is not the place to go into detail, but I want to call attention to a few general points here. First, ageing as a phenomenon is not as intractable as the Medawar view would have us believe. Small changes in a few genes can call off the whole pack of spectres, which no longer consume us from within, but retreat en masse, at least for a period. Second, these rallying effects don't depend on rare or unusual alleles – calorie restriction can produce many of the same benefits, albeit more equivocally in larger animals such as rhesus monkeys or humans. But my point is the same: physiological changes can readily overwrite the seemingly implacable effects of late-acting genes. Third, we see much the same thing in populations of animals shielded from predators, for example on islands. Lifespan is surprisingly plastic: it can double or treble over a handful of generations. The implacable pack of spectres disappears over the horizon. The simplest explanation for this paradox, that ageing is both implacable and flexible, is that lifespan is measured not in chronological time, but in *biological* time. The units, we shall see, should not be tallied in years, but in some measure of flux.

There's another sense in which a genetically deterministic view of ageing falls short: the problem known as 'missing heritability'. The term 'heritability' refers to the proportion of physical differences (known as the 'phenotype') between individuals that can be explained by genetic factors. If the heritability of some trait such as height is 1, then all differences between individuals are explained by genetic variations; if it is 0, then all differences are explained by environmental factors; most traits are a mixture of nature and nurture, giving them a heritability between 0 and 1. Heritability has often been measured through twin studies, on the assumption that identical twins are genetically identical. That's not strictly true, as we'll see, but many complex conditions that run in families, such as epilepsy, might have a heritability in the range of 0.4 to 0.6, perhaps surprisingly low. Even so, if one twin has the condition, then the other twin has about a 50 per cent chance of having it too, regardless of their upbringing. For comparison's sake, the heritability of height is about 0.8.

The problem of missing heritability is more startling in this age of genomics. Genome-wide association studies assign additive risk to multiple genetic variants. If the known heritability of a disease is 0.5 (from twin studies) and the known genetic risk factors account for that full hereditary risk, then together they should add up to an overall risk of 0.5. In fact, known genetic variants often account for less than 10 per cent of the known risk. Nearly all the known heritability of the disease is 'missing' – unaccounted for in the data, not just for epilepsy but for most complex conditions. In part, this is a problem with statistical power. There could be thousands of additional genetic variants, each of which, alone, makes such a trivial contribution to the overall risk that it can be missed altogether in smaller genetic studies; but add them all up and they might account for most of the missing heritability. That may be so. But it's also possible that there is a conceptual problem with the whole idea of heritability. We've seen that the supposedly implacable effects of thousands of late-acting genes can disappear in response to simple physiological changes. Could missing heritability, too, depend in some way on physiology, on some sort of flux state?

The idea is not absurd, for a handful of genes are almost always missing from genome-wide association studies, literally binned on the grounds that they cause more trouble than they're worth. Because there are so few of them, the assumption is that they are statistically trivial. That's a very poor assumption, as these genes have a greater impact on flux state – on our biological age – than any others. What's more, they interact in elusive and unpredictable ways with the rest of the genome. And they alone account for why identical twins are not necessarily genetically identical. They are mitochondrial genes.

Twin pillars of medicine

One voice above all has sung the importance of mitochondria in human health for nearly half a century: Doug Wallace. Wallace trained as a geneticist, completing his PhD in microbiology and human genetics at Yale in 1975. This was already a bold beginning, exploring the resistance of cultured cells to the antibiotic chloramphenicol, which interferes with mitochondria as well as bacteria. Some mitochondria carry an allele (a gene variant) that confers resistance to the antibiotic, allowing the host cell to grow normally. But the strength of resistance doesn't depend only on the mitochondrial gene; it needs to cooperate with the genes in the nucleus too. That cooperation can be undermined by incompatibilities between the mitochondria and nuclear genes. To study these interactions, Wallace made 'cybrids', in which the genes in the nucleus came from one species such as a mouse, while the genes in the mitochondria came from another species, such as a hamster.

Wallace found that antibiotic resistance broke down with genetic distance between the two species crossed. If both the nuclear and mitochondrial genes came from the same species, then the antibiotic resistance worked well; but if they came from distantly related species, such as mice and hamsters, then the cultured cells lost their resistance to the antibiotic. This breakdown in function of crosses between diverging species later came to be seen as a possible mechanism driving speciation; to borrow a phrase, driving the origin of species. A vivid example might be attempts to revive the woolly mammoth from extinction by reconstructing its genome, frozen for millennia in arctic tundra, using the egg cells of an elephant. Asian elephants and mammoths diverged around six million years ago, so they are as closely related to each other as humans and chimps; close, but not close enough for their mitochondria to work well with the nucleus. Even if the complete mammoth genome could be reconstituted successfully, letter by letter, it wouldn't work properly in elephant cells unless the mitochondrial genes were also replaced. That's a taller order. But the bigger point here is that one factor that drove a wedge between elephants and mammoths was the divergence in mitochondrial genes. All of that was implicit, if not yet unpacked, in Wallace's early doctoral work.

Taking a cue from physics, Wallace considered 'energy to be the most important thing'. For any young researcher, setting out in science, choosing your question is perhaps the most important decision of all. What will you do with your life? What do you want to know? Where can you make a real difference? Recalling his own decision, decades later, Wallace's reasoning was exemplary: 'And the thing that made the most energy was mitochondrial; it couldn't be trivial. And if it had its own DNA, it must mutate and that must cause a change in its characteristics and that might cause disease. So I thought it was obvious that one would study mitochondrial energy, and their role in disease.'

Yet even Wallace could scarcely have guessed at the truth back then. Touching on his discoveries is to relive some of the highlights of mitochondrial research. In 1980, Wallace was the first to show that, in humans, mitochondrial DNA is inherited strictly from the mother.[1] Others had already established the

---

1 There are occasional examples of mitochondrial DNA inherited from the

idea of 'uniparental inheritance' of mitochondrial genes (the father's mitochondrial genes are usually not passed on) but Wallace used a spanking new method (cutting up mitochondrial DNA into diagnostic chunks using 'restriction enzymes', which cut in different places depending on the sequence) to demonstrate beyond doubt that children inherit their mitochondrial genes from their mother. He went on to show that mitochondrial genes can indeed mutate, causing diseases such as epilepsy, blindness and neuromuscular degeneration. We now know of at least 300 diseases caused by mutations in mitochondrial genes (plus hundreds of others caused by mutations in the nuclear genes that encode mitochondrial proteins). It's a tragic story, yet just the tip of the iceberg of how mitochondria fashion our lives.

Most mitochondrial mutations are not obviously detrimental. It can be hard to tell. Unlike nuclear genes, which normally come in two copies (one from each parent), mitochondrial genes come in packs, typically hundreds or thousands of copies per cell. A mature oocyte (egg cell) has nearly half a million copies. This means that a mutation, if present in only a few copies, might be masked by numerous normal copies. But it can be much more confusing. The same mitochondrial mutation often produces differing outcomes at different doses, or if set against different nuclear genes, or in distinct tissues, or if interacting with other mitochondrial DNA in the same cell, much of which exhibits tortuously complex behaviour, making mitochondrial diseases elusive and esoteric to this day. For most of us, the complex group dynamics of mitochondria are avoided, at least earlier in life, through what amounts to a clean-up operation

---

father too, very rarely in humans but more commonly in some other species. But these examples are really a footnote to the rule.

in the female germline, again first pointed out by Wallace. This clean-up means that each oocyte is packed with half a million near-identical copies of mitochondrial DNA, very nearly a clone.[2] Most overtly detrimental mutations seem to get sieved out in this process. That's why mitochondrial diseases directly affect only about 1 in 5,000 of us. It's also why mitochondrial genes tend to evolve much faster than nuclear genes; at least ten-fold, and perhaps as much as fifty-fold faster in most animals, including us. Mitochondrial genes are copied far more than nuclear genes, and so they accumulate more mutations. While the most detrimental mitochondrial mutations are eliminated in the germline, plenty of other variations can persist.

The fact that most mitochondrial variation is not obviously detrimental offered some simplified glimpses into human origins and prehistory. By the 1990s, Wallace was tracking the tell-tale trail of mitochondrial variations in human populations back to an ancestral female, now known as 'mitochondrial Eve', the ancestor of all modern human mitochondrial DNA, around 160,000 years ago. The gradual accumulation of differences in

---

2 In his PhD with me and Andrew Pomiankowski, Marco Colnaghi has done some beautiful mathematical modelling on how the female germline selects for the best mitochondria, a process that involves their transfer from multiple germ cells into a single immature oocyte, followed by the programmed death of all the donor germ cells. Overall, this gives rise to about a million oocytes and explains the death of the large majority of primordial germ cells of a growing female fetus before birth. Marco showed that only the selective transfer of mitochondria into primary oocytes could account for the prevalence of mitochondrial diseases found in human populations. Thinking about how selection for mitochondrial quality operates gives striking insights into the developmental architecture of the female germline – why germ cells proliferate to form up to eight million immature oocytes, why most of those die before birth, why oocytes are then effectively put in 'cold storage' for decades, and more. None of this happens in the male germline for the simple reason that males do not pass on their mitochondria. This is the deepest distinction between the sexes, and arguably the reason there are two sexes at all.

mitochondrial DNA between diverging groups can be used to track the migrations of humans out of Africa into Eurasia and eventually to Australia and the Americas. While this gives a two-dimensional view, compared with the richness of whole genomes used today, the broad outlines of the story have been corroborated by genomic and anthropological evidence since then. More controversial is whether these mitochondrial variations confer some level of adaptation to the diverse environments that confronted early humans, from arctic tundra to open savannah, hot deserts and tropical rainforests – a novel form of adaptation that would clearly apply to other species too. The extreme amplification of mitochondrial DNA in the female germline, up to half a million copies per oocyte, means that different oocytes can contain different clonal (or nearly clonal) populations of mitochondria. These give rise to individuals, each of us with our own slightly different mitochondrial DNA, set against our diverse nuclear genomes. Wallace has argued that this mitochondrial variation could aid adaptation to different climates and diets. That makes a great deal of sense, although the evidence remains equivocal.

All of these ideas see mitochondria as somehow different from the rest of the organism, befitting their origin from free-living bacteria two billion years ago. But of course mitochondria are now wholly integrated into their host cells, and their function is central to everything that cells do: not only energy, but also biosynthesis, as stressed in this book. Because energy and biosynthesis necessarily generate a tension at the heart of metabolism, what I've called the yin and yang of the Krebs cycle, mitochondria are barometers of physiological stress, coordinating signalling and even programmed cell death. When should a cell kill itself, if not when it can no longer bind these vital, but disparate, functions together? Not surprisingly, mitochondria play a key role in more complex diseases linked with ageing, diseases from cancer to diabetes and dementia. We considered some of what goes wrong in cancer in the previous chapter, but Wallace's vision goes beyond that, to the very pillars of Western medicine.

Wallace points to Vesalius and Mendel as twin intellectual pillars of medicine. In *De Humani Corporis Fabrica*, the sixteenth-century Flemish physician Vesalius wrote the foundational tome of anatomy, replete with unforgettable plates depicting dramatically posturing figures in various states of musculoskeletal undress, posed in Renaissance landscapes. These plates pack a punch that has echoed down centuries. Vesalius did not fear to pick a fight with the ancients on points of anatomy. Galen had been undisputed for 1,400 years, but Vesalius not only dared to disagree; he was usually right. His legacy can be seen in any hospital today. Go in, and you'll be faced with signs pointing to Neurology, Nephrology, Cardiology, Ophthalmology, Rheumatology, Gastroenterology, Gynaecology, and more. Vesalius would be more at home than most of us, not even put out by the Latin language. Medicine is still grounded in anatomy, to the point that it can blind us to the similarities in conditions that afflict multiple organ systems as we age.

The second pillar of medicine is Mendel, the Augustinian monk who founded genetics – albeit only when his work was rediscovered two decades after his death. Mendelian genetics is textbook stuff. Genes come in pairs, one from each parent. They line up on chromosomes in the nucleus. We inherit different alleles, some of which predispose us to disease. Wallace notes the associated medical corollary: if a clinical trait is transmitted in Mendelian fashion, it is genetic, but if it is not, then the trait must be a consequence of environmental factors.

This corollary is formalised in the concept of heritability and underpins the most sophisticated GWAS today. Except that (as we've seen), these studies usually discard mitochondrial genes, with their non-Mendelian inheritance patterns. So GWAS confounds their effects with the environment, whereas in reality they contribute to the heritability of diseases. Wallace believes we need a more integrated vision, akin to the Chinese *chi*, which translates loosely as energy, or vital force. He frames it beautifully: 'Life is the interplay between structure and energy.'

I would go even further. Mitochondria are flux capacitors. Not invented by Doc Brown and fitted in a time-travelling DeLorean, but quite literally. Mitochondria convert flux through the Krebs cycle into electrical membrane potential. The membrane is a capacitor: a thin insulating layer that separates two electrically charged aqueous phases, generating powerful electrical fields across the membrane. Changes in metabolic flux can increase or decrease this electrical forcefield, affecting everything from ATP synthesis to the flux of ROS or the synthesis of that powerful reductant, NADPH. So electrical membrane potential has a dynamic range, singing like a string on a violin. Electrical fields resonate out from the membrane, capturing the nearby jostling crowd of molecules in an oscillating dance. These resonant oscillations rise in crescendo or fall in diminuendo as the membrane potential strengthens or weakens. The dynamic range depends on flux, but also feeds back, shaping forward or reverse flux through the Krebs cycle. As we saw in Chapter 5, if the electrical potential is too high (because ATP is not being consumed) then Krebs-cycle flux must slow or even reverse. That signals directly to the nucleus, through succinate and citrate, controlling the activity of thousands of genes, the epigenetic state of the cell. There is a subtle and continuous interplay between the singing membrane potential, flux through the Krebs cycle and the epigenetic state of the cell. To my mind, even slight differences in flux capacitance between individuals could explain why we age in myriad ways.

In praise of inbreeding (in flies)

How can we distinguish between the genetic contribution of the nucleus and the mitochondria? Just as studies with identical twins have given an indication of the heritability of complex diseases, so too 'genetically identical' flies give a profound insight into the role of mitochondrial flux.

I'm talking about *Drosophila melanogaster*, the tiny fruit fly (or technically, vinegar fly) that has been the source of grand insights into genetics, going back to the earliest days of the discipline when Hermann Muller first showed that X-rays can induce mutations in *Drosophila* genes. Muller also discovered a clever trick that enabled him to preserve interesting mutations – what he termed 'balancer' chromosomes. These chromosomes contain multiple inverted sequences, meaning that big chunks have been cut out, flipped around and reinserted backwards. That suppresses recombination during sex and makes it far more likely that all the genes on the chromosome will stay together in exactly the same order. Just to make sure, balancer chromosomes also contain gene variants that confer specific traits, such as hairy backs or curly wings, which can easily be recognised in crosses, proving that the genes are still linked together. And then, as a final failsafe, the flies are inbred to ensure they really are as close to being genetically identical as it is possible to be. You can imagine that it's feasible to do these genetic tricks in tiny flies, which live for a couple of months and have brains the size of pinheads, when it could not be justified ethically in larger animals. Ethical regulations don't even classify flies as animals, though those who work with them become fond of their antics. Male flies will dance the conga in chasing females, mostly at dawn or dusk, and then sleep it off during the day. And fruit flies love to get drunk; they live on rotting fruit, gaining their proteins from the yeast that ferments the fruit to fly-wine. I confess to quite a strong fellow feeling.

I knew little of this until Flo Camus arrived in the lab next door to mine at University College London. She had done her PhD in fly genetics and mitochondrial function in Melbourne and was now full of ideas and plans – and as excited about mitochondria as I am. Flo was particularly interested in a problem known as 'mother's curse', and we began to work together on the question. The problem here is precisely that mitochondria are inherited from the mother only. My own mitochondria are an evolutionary dead-end, so my sons inherited all their mitochondria from their mother. That pattern goes back to mitochondrial Eve and beyond; possibly right back to the first complex cells nearly two billion years ago.[3] The upshot is that mitochondrial DNA is selected for how well it works in females, but not in males. It doesn't matter how good or bad my mitochondrial DNA might be for me, it's going nowhere. If there are no differences between males and females in metabolic demands, then male fitness would be the same as female

---

3 This is a fine example of how solving one evolutionary problem leads to another. Uniparental inheritance made it possible to select for mitochondrial genes: they are always inherited in populations, so it is important to make this population as clonal as possible. In animals, that requires careful selection in the female germline; mixing this high-quality clonal population with random mitochondria from the father is a very bad idea. Sperm therefore contain a minimal contingent of mitochondria, which are normally excluded from the zygote (the fertilised egg). Uniparental inheritance thus solves the problem of selection for mitochondrial quality, but sets up its own contingent problem – mother's curse.

fitness, and maternal inheritance would not be a problem. But that's not the case. On average, men have a resting metabolic rate about 20 per cent higher than women, which is linked with a slightly greater tendency to burn carbohydrates rather than fats, and accordingly, different proportions of muscle and bodily fat. Equivalent differences between sexes are normal across most species. This means that some mutations could in principle be neutral, or even beneficial, in females, but bad for males. That notion is supported by the fact that men (or rather, boys) are more likely to suffer from mitochondrial diseases than women. This is consistent with the idea of mother's curse – males inherit mitochondrial DNA that has been optimised for its performance in females, making it more likely to go wrong in men. Some even argue that's why women generally outlive men.

Yet mitochondrial diseases are uncommon, and the difference between the sexes is real but not stark. Selection has found a way to compensate: genes in the nucleus can improve male performance. These have been tailored over many generations to 'fix' problems with mitochondrial DNA in males. For this form of compensation to work, specific alleles in the nucleus must counteract randomly accumulating mutations in mitochondrial DNA, and that in turn requires that they be inherited together – that they coexist in the same population. If so, then outbreeding to very different populations (which lack these compensating nuclear genes) could potentially unmask mother's curse. But where should we draw the line? Crosses between species do indeed cause 'hybrid breakdown', attributed to incompatibilities between mitochondrial and nuclear genes. So what about crosses between divergent populations within the same species? Falling in love with people from other cultures is as human as we can get, and I will always celebrate it. But we ought to know if hybrid vigour can be undermined by a higher risk of mitochondrial diseases. Spoiler alert: I don't think so. But there's plenty else to ponder on here.

This was the starting point for Flo's work. She had bred a number of different fly lines using the balancer chromosomes. Each of these fly lines has a particular mitochondrial genome, derived from natural populations around the world (it's important to emphasise these are not detrimental mutations, simply variations found in normal populations) set against an identical nuclear genome. In other words, the flies are like identical twins – genetically identical in the nucleus, but differing in their mitochondrial DNA. That allows us to mismatch the mitochondria to the nuclear background: to study how mitochondrial genes interact with nuclear genes by 'fixing' the nuclear background so it is no longer a variable. In this case, we were interested in mother's curse. Presumably, if nuclear genes compensate for specific mitochondrial variations, then switching the mitochondrial DNA should unmask the curse, giving males new problems. The most likely problem, as others had reported previously, was male infertility, or at least reduced fertility. And some of our mismatched lines did indeed show lowered male fertility, as expected.

You may be wondering what all this has to do with ageing. We hadn't set out to study ageing at all, but we were taken aback by our initial findings – always a good sign in science. Isaac Asimov (a biochemist, by the way) once said that new discoveries are not accompanied by 'Eureka!', but 'That's funny ...'. In this case, we'd measured the rate of respiration in various tissues, as well as the proportion of oxygen that picked up stray electrons from the respiratory complexes to form ROS, otherwise called 'free radicals'. We were using an Oroboros respirometer to make these measurements, and that allowed us to determine the speed at which ROS were forming during respiration in real time; what's called ROS flux. One of the mismatched fly lines with low male fertility seemed to have slightly higher ROS flux in its testes, coming from complex I of the respiratory chain. We got the picture, and stored all our data in a folder entitled 'Stressed testes'. Our first (unoriginal) conclusion was that mismatched mitochondria caused male infertility by stressing the testes.

The cure seemed obvious: lower the stress. Give the flies an antioxidant. We tried one called NAC (for N-acetylcysteine), but it didn't affect the males much. *Yet it killed most of the females in one fly line!* Just one line. The other lines looked a bit ropey but seemed OK. We didn't say 'That's funny'; it was anything but. It was quite shocking. Exactly the kind of unexpected finding that leads to places in science. Giving the flies an antioxidant had little effect on the males but killed females with one type of mitochondria while sparing the others. Remember that these were all genetically identical flies, except for the mitochondrial DNA, which differed in just a few SNPs.

To generalise: the response to drugs (an antioxidant in this case) can vary dramatically, depending on a few tiny differences in mitochondrial DNA, with big differences in outcome between males and females. Flies have exactly the same mitochondrial genes that we do, and these are inherited in the same way, so the same principles apply to us too. I'm going to focus on NAC for the time being, because it can tell us a lot about the possible mechanisms of ageing too, but don't think it's an exception. We've found big differences in outcome with other treatments too, including high-protein diets. If you ever wondered why some diets work well for some people but not for others, this is part of the answer: subtle differences in mitochondrial function can produce very different responses to the same treatment in flies, and most likely in humans too. I say subtle differences.

In our fly lines we found no link between the number of SNPs difference in mitochondrial DNA and outcome – a handful of differences could give a catastrophic outcome, while a difference of fifty SNPs might have no effect. The issue was not with the number of SNPs, but a few unlucky interactions with some nuclear genes. Taking a bigger picture, this means that particular differences between individuals are more important than average differences between groups. 'Race' has no effect.

Bringing balance to the force

So what was killing the female flies? We don't know for sure, but we did measure one big difference between the males and females, and between the female line that died versus those that seemed fine. The antioxidant NAC suppressed respiration, especially at complex I of the respiratory chain. The respiratory rate in the vulnerable fly line was barely *a third* that of controls, and half that of the females in surviving lines. In effect, the antioxidant was suffocating them internally. What was going on?

The answer seems to be linked to the flux of ROS. A really striking finding was that ROS flux hardly changed, least of all in the powerful flight muscles of the thorax – it remained more or less constant, despite the suppression of respiration by NAC. That's surprising, because ROS are produced by respiration, so suppressing respiration should have some effect on ROS flux, yet there was no change. A much lower rate of respiration was producing the same overall rate of ROS flux. But could it actually be the other way round, we wondered? Was respiration being suppressed precisely to keep ROS flux within normal bounds?

If the electrical potential on the mitochondrial membrane is a high-energy force-field, then that ominous electrical crackle is ROS flux. As a rule of thumb, the higher the potential, the greater the ROS flux, the more scope for damage to the surrounding materials. Balancing the force means preventing a dangerous system overload; lower the voltage before it blows. If the system becomes dangerously overheated, then it might be best to suppress respiration, to lower the voltage. And if the system becomes damaged and overheats even at a normal voltage, then you'd better suppress respiration even further, down to very low voltages. On the face of it that's sensible, yet in the case of our flies, respiration was being suppressed to the point of death. That sounds like madness.

But it's not out of the question that keeping ROS flux within normal bounds could be more important than respiration specifically through complex I. There are other ways of generating ATP, as we saw in relation to cancer. And the male flies weren't doing so badly, despite having the same mitochondrial DNA. But that loophole just brings us back round to the same question: if it's possible to make do, despite suppressed respiration at complex I, what was killing the female flies?

NAC has a specific effect as an antioxidant, which is to raise the levels of a critical antioxidant molecule, the sulfurous little peptide glutathione. Supplementing the diet of fruit flies with NAC raised glutathione levels in all tissues, which is usually taken to be a good thing. Yet those rising glutathione levels were apparently killing some fly lines and not others. The only difference between these lines lay in their mitochondrial genes, so the primary deficit had to be in respiration; giving the flies NAC amplified the differences between the lines. To understand what was going on, and why this is relevant to ageing, we need to think about what glutathione and ROS actually do in cells. But first, before we go down this rabbit hole, let me remind you, from the previous chapter on cancer, that suppressing respiration doesn't only lower energy availability; it hinders Krebs-cycle flux, which exerts downstream effects on the activity of thousands of genes. By suppressing respiration, NAC *has* to be having serious effects on the epigenetic state of tissues: which genes are switched on or off. Enrique Rodriguez and Gla Inwongwan in my lab are looking into that right now, so I can't tell you any more details yet, but I can give you a sense of what we think might be going on.

The normal (reduced) form of glutathione, GSH, contains a thiol group – a sulfur bound to a hydrogen. This thiol can pass on its H to other molecules including ROS such as hydrogen peroxide (H_2O_2), giving water plus oxidised glutathione. When oxidised, glutathione can bind to itself through a disulfide bond to form what's called GSSG:

$$2GSH + H_2O_2 \longrightarrow GSSG + 2H_2O$$

Notice here that each glutathione, GSH, passes on a single H to H_2O_2 to form two water molecules, plus a single oxidised glutathione, the composite GSSG. The fact that each glutathione trades with a single H enables it to intervene directly in free-radical reactions, which by definition involve single unpaired electrons. It can also pass on its single H to regenerate other cellular antioxidants such as vitamin C and vitamin E. For this reason, glutathione is often seen as the master antioxidant in cells.

To recycle the GSSG back into two GSH requires 2H. These 2H come from the powerful reducing agent NADPH (you might remember that in Chapter 5, I suggested you think of the P as standing for 'power', to discriminate it from the weaker NADH). So now there's a second factor to consider here: if a cell can't make enough NADPH, then it won't be able to regenerate all the GSSG back into GSH. There are only a few ways of making NADPH; we previously touched on the pentose phosphate pathway, the malic enzyme and the mitochondrial transhydrogenase. Either directly or indirectly, all these pathways depend on Krebs-cycle flux. So we can see straightaway that a deficit in respiration which compromises Krebs-cycle flux could in turn affect the regeneration of GSH – a greater proportion remains oxidised as GSSG. The glutathione pool is therefore a sensitive indicator of the health of the cell. If it becomes more oxidised (so there is more GSSG relative to GSH), that sends a powerful signal: all is not well with the basic metabolism which underpins everything that cells do. There's a sickness at heart. We were seeing exactly this in the flies that were dying. Their glutathione pool was more oxidised – a slightly greater proportion was in the form GSSG in the flies that died.

There are more layers. GSH can also regenerate thiols in proteins. As with GSSG, oxidised thiols in proteins tend to form disulfide bridges. By linking together different bits of a protein, disulfide bridges change the structure and function of proteins, activating or deactivating them in ways that are still imperfectly understood. You could think of this as a simple on–off switch for a protein. But there might be hundreds or thousands of copies of the same protein in a cell, each of which can be in the 'on' or 'off' state, giving rise to a weighted signal, like a show of hands to make a direct democratic decision. In this case, the show of hands depends on the availability of GSH (which converts disulfide bridges in proteins back into thiols). GSH behaves like an infiltrator in a crowd, whispering honeyed words to go home and not start a fight. And once again – the availability of GSH depends on the regenerative power of NADPH, and so ultimately on Krebs-cycle flux.

There's a final nuance I need to mention. Oxidised glutathione doesn't only take the form GSSG; there can be a combination of the two processes I've mentioned, whereby an oxidised thiol from glutathione binds to an oxidised thiol in a protein, forming a disulfide bridge that binds glutathione itself to the protein – what's known as 'S-glutathionylation'. That amounts to another signal, again signifying that all is not well with the world of the cell. I mention S-glutathionylation because one of its known effects is to suppress respiration. That's probably what was happening in our flies, although we've not proved that yet. But it does make sense. Under normal conditions, most ROS flux seems to come from the FeS clusters in complex I. When electron transfer is sluggish, electrons can escape from these clusters and react directly with oxygen, forming more superoxide and hydrogen peroxide than usual; that's to say, there's increased ROS flux. If all else fails, the fallback solution to this problem is to restrict the total number of reactive FeS clusters that can react with oxygen directly – make less complex I. And that's what S-glutathionylation seems to do.[4] The price is serious: it suppresses the most efficient form

---

4 Biology is always more complex than you'd like it to be. It took me a long time to see this relationship, because I already had a hypothesis in mind, which I have written about in previous books, and which has some experimental backing: increased ROS flux signals locally (to the mitochondrial genes) that respiratory capacity is insufficient. The response is to make more components of the respiratory chain – increased respiratory capacity should then fix the problem. Here I'm saying the opposite. Can they both be true? Yes, I now realise; it's just a matter of timing. Increasing respiratory capacity will only fix the problem if the problem is strictly one of capacity. If respiration is in fact partly broken – for example, because of mitochondrial mutations – then increasing capacity won't help much. There will still be sluggish electron transfer, and a high rate of ROS flux. That will gradually oxidise the glutathione pool, not over minutes, but over extended periods – perhaps in our case, many years. Oxidation of the glutathione pool leads to S-glutathionylation of proteins, which suppresses

of respiration, the form that pumps ten protons for every 2H, instead of just six. That's quite a hit.

Let's take stock with our fruit flies. We know that NAC suppressed respiration in some fly lines, but not others. We know that glutathione levels increased in all fly tissues, so NAC was doing what it is supposed to do – promoting the synthesis of glutathione. We know that the glutathione pool was more oxidised in the fly lines that died. We know that oxidation of glutathione signals the suppression of respiration at complex I, which in principle decreases ROS flux. We know that ROS flux was in fact unchanged in muscle tissue, so presumably the suppression of complex I did indeed bring elevated ROS flux back within a normal range. We know that the only difference between our fly lines lay in their mitochondrial DNA, so the primary deficit had to be in respiration. We know that any deficit in respiration necessarily disturbs Krebs-cycle flux, because the Krebs cycle cannot spin if NADH is not being oxidised in respiration. Finally, we know that regeneration of reduced glutathione (GSH) requires the powerful reductant NADPH, which itself depends on Krebs-cycle flux. So we can infer that a primary deficit in respiration hindered the regeneration of GSH, signalling a ROS problem that flies fix by suppressing respiration further – a dangerous vicious cycle.

In sum: tiny differences in respiration between fly lines were amplified by the antioxidant NAC. The flies kept ROS flux within normal bounds by suppressing respiration at complex I, in some cases to the point of death. In the most vulnerable fly line, the males seemed fine, but most females died. Why?

---

respiration at complex I. That fixes the ROS flux problem more permanently, as there are now fewer reactive FeS clusters in the cell. But of course it also limits respiratory capacity. Ironically, all this was implicit in a figure I drew in a 2011 paper, but only now do I understand my own figure properly.

I suspect because the females had more constraints on their Krebs-cycle flux, probably because egg production in female flies is a metabolically costly business. Suppression of respiration was more costly for them than for males. In any case, the greater the stress (relative to demands), the greater the respiratory suppression, and the more severe the cost – ultimately death. I'm telling you all this because I could write almost exactly the same thing about ageing.

The age-old problem

I suggested at the beginning of this chapter that ageing is not linked to chronological time, but to some sort of biological time measured in flux. You may be familiar with the simplest version of this idea, that our lives are measured in heartbeats, with all animals having roughly the same number of heartbeats in their life, around a billion for both the pygmy shew (squeezed into a year and a half, at an extraordinary rate of 1,300 beats per minute) and the elephant, with its ponderous twenty-eight beats per minute for around seventy years. You'll be pleased to know that humans are a modest exception to this rule of thumb, clocking up more than two billion heartbeats at a rate of about sixty per minute over seventy years.

There's at least a grain of truth in this idea, although it won't bear out the wag who quipped 'If we have a finite number of heartbeats, I don't intend to use up any of mine running around doing exercises.' If they had done so, they would likely have lived a little longer (albeit their heart rate would have slowed down too). My point is that there is a broad correlation between metabolic rate (the rate of living) and lifespan. The faster we burn through resources, the faster we burn out. Small animals typically have faster metabolic rates and shorter lifespans; large animals tend to have lower metabolic rates and live longer.[5] While there are plenty of exceptions to this rather loose rule, there are also simple explanations for most of them – given any particular metabolic rate, animals might invest more or less in damage limitation. The devil is in the detail here. This is an old idea, going back at least a century, and many clear predictions turned out to be wrong. For example, it's not true that damage accumulates in the form of mutations in DNA as we age. Mutations certainly do accumulate, but as we saw with cancer in the previous chapter, not quickly enough to drive ageing. A variant of the mutation accumulation idea relates to mutations specifically in mitochondrial DNA, but they too do not correlate meaningfully with the rate of ageing.[6] The 'free-radical theory of ageing' is also problematic. The idea is simple enough, but the evidence is equivocal, to put it mildly. Free radicals (the same ROS escaping from respiration that we've been talking about) attack nearby membranes, proteins

---

5 Quite why larger animals usually have a lower metabolic rate is surprisingly abstruse. In part, it relates to the rate of heat loss, which scales with the surface area of animals. An elephant-sized pile of mice produces about twenty-one times more heat than an elephant does; if the elephant produced that much heat, it would literally melt. It may partly relate to the fractal geometry of supply networks, as argued by Geoffrey West and others (but see my earlier book *Power, Sex, Suicide* for a critique of this idea). And it partly relates to the economies of size. The great Canadian comparative biochemist Peter Hochachka showed that the demands on visceral organs such as the liver fall with body size, for reasons that are not still not fully explained.

6 'Mutator mice' have an error-prone mitochondrial DNA polymerase that introduces mutations into mitochondrial DNA. Double mutants (in which both copies of the gene are error prone) have very high numbers of mitochondrial mutations. They age rapidly and die young. But mice with a single copy of the error-prone polymerase are normal, despite accumulating a heavy burden of mitochondrial mutations: plainly, the number of mitochondrial mutations does not correlate well with the rate of ageing. These mice also had quite normal rates of ROS flux.

and DNA, the theory goes, so antioxidants should hinder that damage and prolong lifespan. But there's zero evidence that antioxidants can prolong lifespan, and in many studies the very opposite happens: they turn out to shorten lifespan, whereas pro-oxidants (which supposedly exacerbate damage) actually prolong lifespan. As originally formulated, the free-radical theory of ageing is wrong.

But there may be truth in more subtle interpretations, which consider the role of signalling and endogenous antioxidant enzymes. While ROS *can* cause damage, the lurid devastation often attributed to them is exaggerated, while their more subtle physiological role has been harder to understand. Damage happens in many other ways too, such as proteins unfolding, linking to each other, or simply accumulating too much (what's known as 'hyperfunction'). It costs time and energy to build the perfect protein, and there are tens of millions of them in each cell. If either time or energy is limited (and the faster you live, the less you will have of each to spare), then it's only a matter of time before some proportion becomes dysfunctional. The question is, does that matter? The answer depends a great deal on the context. How long do you expect to live anyway? What proportion of your resources do you need to invest in reproduction or competing for a mate? How much do you want to spend on cleaning up the mess or preventing the mess from building up in the first place? The contrasting answers to these questions explain why lifespan only correlates loosely with metabolic rate – but to ignore metabolic rate altogether, as somehow old-fashioned and genetically nebulous, is to ignore perhaps the single most important determinant of longevity: how quickly the cellular machinery needs to be built, how heavily it must be used, and how soon it has to be replaced. The axiom 'live fast, die young' is far from an immutable law, but it is a harsh thermodynamic reality that cannot be evaded easily. Free radicals contribute to the overall rate of damage, but not necessarily all that much.

The problem is that we have focused too much on the colourful chemical reactivity of free radicals, and too little on their essential physiological role. ROS are such vital physiological signals that cells go out of their way to keep ROS flux within tight physiological limits. Redox tone – the balance of electron sources and sinks in a cell – is as critical to homeostasis (our normal chemical balance) as temperature or acidity. These limits may be temporarily exceeded in infections or illness, as part of the stress response linked to inflammation and immune activation, just as temperature rises with a fever. But in both cases the normal bounds are restored as soon as possible. Remember our study with flies, which showed that ROS flux barely changed despite – or rather, *because of* – the suppression of respiration. Redox tone is ultimately regulated by fiddling with the knobs that control respiration.

We saw in Chapter 5 that suppression of respiration at complex I slows down the oxidation of NADH, promoting reverse flux through much of the Krebs cycle. Krebs-cycle intermediates such as succinate and citrate begin to leak from the mitochondria into the cytosol, where they can stabilise transcription factors such as HIF_{1a}, switching on genes that promote growth and inflammation – and tragically, as we age, cancer. This amounts to an epigenetic shift in the profile of gene activity linked with older age, a 'senescent state'. Some researchers see this as a 'quasi-program' that runs on past its sell-by date. I see that as an over-interpretation – the same pattern is explained by no more than respiratory suppression with age. What we didn't answer in Chapter 5 was exactly *why* respiration is suppressed with age.

The most reasonable broad-brushstroke answer is surely some level of damage, whether that damage is caused by protein unfolding or cross-linking, oxidation by ROS or glycation (the tendency of sugars such as glucose to react with proteins and lipids, decorating them with 'sticky' tails). Presumably, any incremental damage does not impede respiration or Krebs-cycle flux much when we're strutting our stuff in our reproductive prime. Respiration is efficient, so we can drive both biosynthesis and ATP synthesis through the normal oxidative Krebs cycle. To do so, we need to extract carbon skeletons from the Krebs cycle for the synthesis of amino acids, fatty acids, sugars, nucleotides and more: the cycle is in 'roundabout' mode, with traffic entering and exiting at every junction, albeit with overall flux operating in a forwards direction, like a normal roundabout. To power ATP synthesis, some intermediates will need to spin through the full cycle. But now think about what that entails. Let's say we make citrate from acetyl CoA and oxaloacetate, the standard first step of the Krebs cycle. Then we bleed off some citrate to the cytosol to synthesise fatty acids. The effect of this is to lower the concentration of citrate relative to later intermediates such as α-ketoglutarate. The same principle applies at every step – whenever an intermediate is bled off for biosynthesis. The more intermediates that are bled off, relative to those completing the cycle, the flatter will be the concentration difference between those remaining, and the more likely it is that flux will grind to a halt or tip into reverse. It is intrinsically difficult to do both biosynthesis and ATP synthesis at once.[7]

---

7 It might seem that this problem could be solved by upregulating 'anaplerotic' pathways – adding in specific intermediates, modifying them through part of the Krebs cycle, and then removing them again. In other words, the cycle is acting as a roundabout rather than a full cycle, with flux coming in at one

Forward flux also depends on the concentration of other reactants, most notably NAD^+ and NADH. To go from citrate to α-ketoglutarate requires 2H being stripped out and passed to NAD^+, to form NADH. So long as there is plenty of NAD^+ available, and not too much NADH, then everything is fine – flux will keep on going forwards. But to keep this NAD^+–NADH balance means transferring the 2H on from NADH into the respiratory chain, to burn in respiration, regenerating the NAD^+. That operation depends specifically on complex I of the respiratory chain, which strips 2H from NADH, to regenerate NAD^+. You can imagine the problem now if complex I activity is suppressed, to maintain ROS flux within normal bounds. NADH is no longer oxidised as effectively, and so begins to build up. That favours reverse flux, in this case from α-ketoglutarate to citrate. In other words, forwards flux through the Krebs cycle can supply all the carbon skeletons and ATP needed for growth, but only when respiration is efficient. When we're young.

I suspect this difficult trade-off explains in part the 'handicap principle' – the idea that sexual ornaments such as the peacock's tail or the amazing eyes-on-stalks of the imaginatively named stalk-eye fly rests on the relative allocation of resources to growth or energy: the fittest individuals have a respiratory system that is capable of doing both at once, with the Krebs cycle in full forwards mode much of the time, whereas less fit individuals can't sustain both biosynthesis and energy

---

junction and leaving at another. In fact, that just exacerbates the problem. Look back at the figure on page 222. Glutamine feeds into α-ketoglutarate, which steepens the concentration difference to succinyl CoA, but also flattens the difference between α-ketoglutarate and citrate. That tends to push reverse flux to citrate, as also depicted in the figure. And in fact, work from Sarah-Maria Fendt and colleagues shows that the ratio between concentrations of α-ketoglutarate and citrate really does determine the direction of flux through the cycle, whether forwards, or in reverse to citrate.

metabolism by forwards flux, so must make a choice. Their stunted tail or narrow eye-span or less vibrant colour betrays a Krebs-cycle flux compromised by deficient interactions between the proteins encoded by mitochondrial and nuclear genes. That makes the signal genuinely honest; nothing less than optimal respiration can balance energy with growth.

Any damage to the respiratory chain will tend to increase ROS flux, most of which derives from the iron–sulfur clusters of complex I. And ultimately, the only way for animals to get this back under control is to suppress complex I.[8] The problem is that suppressing complex I slows down the oxidation of NADH in mitochondria, promoting reverse Krebs-cycle flux and an epigenetic shift linked to growth and general inflammation, as we saw with cancer – a senescent flux pattern. More ATP is formed by fermentation, with electrons being fed into alternative sinks such as complex III (via the glycerol phosphate shunt), enabling the pumping of six rather than ten protons for each 2H oxidised. Overall, we have less energy, tend to gain in weight, find it harder to burst into explosive action and suffer from chronic low-grade inflammation; aches and pains. Ageing, eh! On the positive side, this diminished state might be stable for decades. But notice how far removed this is from the venerable free-radical theory of ageing. ROS flux doesn't rise sharply with age, in some crescendo of oxidative damage. On the contrary, it barely changes – we suppress our own cell respiration instead, slowing down gently and losing function ... suffocating inwardly.

---

8 Plants and simple (tiny) animals don't need to suppress complex I, as they can short-circuit the respiratory chain (using an alternative oxidase or uncoupling proteins) to generate heat, rather than ATP. Dissipation of heat in this way swiftly oxidises NADH and lowers ROS flux, speeding Krebs-cycle flux. But overheating is a delicate business for most animals, and is more likely to induce an untimely demise than control ROS flux.

We're all individuals

Ageing reminds me of that absurd scene in Monty Python's *Life of Brian*, where Brian admonishes the crowd that they don't need to follow anyone – they're all individuals. The crowd chants back, 'Yes, we're all individuals!' apart from one lone, doleful voice who says 'I'm not.' Likewise, we all get older in much the same way; we're all part of the crowd. Yet as I noted at the beginning of this chapter, we each succumb to different ailments with age, whether it be cancer, Alzheimer's disease or some heart condition. Take your pick. We're all individuals. But the spectre of missing heritability means that only a small proportion of the established genetic risk for these conditions is explained by known variations in nuclear genes. So how exactly are age-related diseases linked with ageing?

Our flies illuminate part of the answer. We all – even identical twins – have a unique combination of mitochondrial genes set against our nuclear genomes. The flies show us that seemingly trivial incompatibilities between these two genomes can produce devastatingly different outcomes, with the antioxidant NAC killing some flies but not others, depending on their mitochondrial genome, even though they were all identical twins in their nuclear genes. NAC produces a stress that is not dissimilar to ageing: it, too, suppresses respiration at complex I, especially in flies with a tendency to produce slightly more ROS in the first place, those that happen to have some minor incompatibility between their genomes. Slightly mismatched genomes will tend to slow down electron transfer to oxygen, promoting the same epigenetic shift that characterises the hypoxic (low oxygen) response in animals dealing with hydrogen sulfide in stagnant muds, or infections and inflammation, and in the end, ageing and cancer. Mismatched mitochondria simulate suffocation.

Subtle mitonuclear mismatches probably have little effect

in youth but can be amplified by damage or stress as we age. The common denominator is suppressed respiration, altered Krebs-cycle flux and an epigenetic shift, accelerating ageing. What really happens depends on many other factors, including use and diet. Using your mitochondria – aerobic exercise – doesn't wear them out faster. On the contrary, it promotes their turnover and regeneration, obliging us to dedicate more resources to repairing any damage done. Idleness has the opposite effect. Overeating is not good for our mitochondria either. Remember that the biochemical pathways promoting growth, in cancer or sexual maturation, are closed down by too much ATP – instead, we need carbon skeletons from the Krebs cycle, and that powerful driver of biosynthesis, NADPH (what I called 'power hydrogen'). If we sit around and overeat, the aerobic capacity of our mitochondria dwindles, hastening the ravages of ageing. The good news is that we can improve our health – we can make ourselves more youthful in our biological age, even if never young again, through improving our lifestyle.

But however well we live, we can still be undone by unlucky genetic variants. Those variants do not depend on our chronological age, but rather our biological age, and specifically on how these variants respond to the dwindling availability of ATP and the shifting pattern of gene activation in the nucleus, tissue by tissue. By their very nature, different tissues express different genes, with the effects of each SNP depending on which genes are active, overlaid on the energetic and biosynthetic demands of each tissue, and the ability of our weakening mitochondria to meet those needs. This in turn depends on how our mitochondrial genes interact with the unique set of nuclear genes that are expressed in each tissue. Remarkably, the protein content of mitochondria differs by about a quarter from tissue to tissue, so the same mitochondrial genes have to operate in radically different genetic environments. And that brings us back around to a point I made in Chapter 4 about the 'symbiosis' between tissues in animals. Diverse tissues support each other's function through complementary patterns of Krebs-cycle flux. Cancer cells selfishly exploit this delicate balance to their own ends (and ultimately, that's why cancers kill), but most tissues don't have this flexibility. They depend entirely on their metabolic needs being serviced by other tissues, and have little scope to change their own Krebs-cycle flux with age.

Just think about the brain's addiction to glucose. Although the brain can make do, or even thrive, on other substrates (such as ketones) it much prefers glucose. Positron emission tomography can be used to study metabolic processes. Brightly coloured PET scans depict regions of the brain 'lighting up' as a radioactive form of glucose floods in, and neuronal networks power up. But why glucose? The heart does perfectly well most of the time burning fatty acids (which pack in more calories per gram), only switching to glucose in times of stress. Yet the brain eschews fatty acids or amino acids as energy sources under most circumstances. The reason probably relates to the brain's need for sudden shifts in power, as neurons fire up. They need to maximise the electrical potential of neural mitochondria.

To understand why membrane potential matters, consider another tissue with an equivalent addiction to glucose: the beta cells in the islets of the pancreas. They're addicted to glucose because their job is to detect high blood glucose levels (usually after a meal) and secrete the hormone insulin in response. Insulin then promotes the uptake and metabolism of glucose around the body. Did you ever wonder how the beta cells detect high blood glucose? The answer is exquisite: *insulin secretion depends on mitochondrial membrane potential.* The more glucose there is, the higher the membrane potential rises, and the greater the secretion of insulin. Mitochondria really are flux capacitors. Now you see the problem with diabetes. Damage to the machinery of respiration in the islet mitochondria – damage that's typically caused by the reactivity of glucose itself – lowers the membrane potentials achievable, which saps insulin secretion. As secreted insulin levels fall, glucose is taken up less easily by the brain, despite high levels in the bloodstream. Perpetually high blood glucose drives a vicious cycle, in which tissues eventually become resistant to insulin. And that gives the brain a particularly serious problem, as it relies so heavily on glucose. It can't simply revert to other fuels or alternative Krebs-cycle fluxes, like a cancer cell. Instead it must crank on as normal, despite the grinding gears. So the harder it is to respire glucose, the greater the risk of disease. Accordingly, people with type II diabetes have double the risk of Alzheimer's.

In a pioneering reconception of the pathology underlying Alzheimer's disease, Estela Area-Gomez and Eric Schon at Columbia University, New York, have linked neuronal degeneration with the damage accruing to cellular compartments known as mitochondria-associated membranes or MAMs.[9] Perhaps first among their tasks, MAMs are the gatekeepers for calcium ions (Ca^{2+}). When the MAMs open their calcium

---

9 These are part of an extensive internal membrane system in cells called the endoplasmic reticulum, which is responsible for synthesising, packaging, folding and transporting proteins and other molecules such as lipids around the cell. The endoplasmic reticulum also plays an important role in sequestering ions such as calcium (Ca^{2+}) which can be released into the cytosol in response to signals such as hormones. In places, the endoplasmic reticulum comes into close apposition with the mitochondria. Here they are held in close contact by protein complexes that span between the membranes. These are the MAMs, which communicate with mitochondria in ways only now being explored in relation to Alzheimer's disease.

gates, Ca^{2+} ions flood into mitochondria and ramp up the activity of the pacemaker enzyme, pyruvate dehydrogenase. This enzyme controls flux through the Krebs cycle, and by extension, mitochondrial membrane potential. Here's how it works.

Neurons need to accelerate from zero to sixty in a split second when called upon to fire. Burning glucose maximises the speed and efficiency of ATP synthesis through the classic textbook pathway. Glucose is swiftly broken down to form pyruvate, which is then clipped of CO_2 and 2H by pyruvate dehydrogenase to give acetyl CoA. That is fed into the Krebs cycle to feed the furnace. This whole system is powered up almost instantaneously as soon as neurons need to fire. The acceleration is achieved by the flood of Ca^{2+} ions from the MAMs, activating pyruvate dehydrogenase. The Krebs cycle leaps forward, stripping out 2H and pressing it into the respiratory chain, mostly at complex I, pumping protons and powering up the electrical membrane potential. When Ca^{2+} ions are present, the membrane potential rises steeply, and the rate of ATP synthesis swiftly doubles. The speed of this process is plainly useful if you just spotted a tiger behind that bush, and crucially draws on the full dynamic range of mitochondrial membrane potential, powering up from idling mode, around 120 mV to more than 160 mV in seconds. We'll consider how this relates to consciousness in the Epilogue, but for now let's just see what goes wrong in Alzheimer's disease.

Think about the problem here. To fire normally, neurons must take up glucose, but insulin resistance means that they struggle to do so.[10] The MAMs attempt to compensate for this

---

10 High glucose in the blood (hyperglycaemia) suppresses the uptake of glucose by neurons, via mechanisms that are not well understood (though glucose

deficit by pumping more Ca^{2+} into the mitochondria, activating pyruvate dehydrogenase further and rescuing neuronal function. But excess Ca^{2+} is itself damaging, and the MAMs become swollen and damaged over time. MAM dysfunction is linked to many other features of Alzheimer's in an uncanny way. Just to give a single example, the amyloid protein that forms plaques in the brains of Alzheimer's patients derives from a longer precursor protein which is processed in the MAMs. When the MAMs become damaged, faulty processing of the precursor protein drives the formation of plaques. Other proteins and lipids linked with Alzheimer's disease depend on MAM activity in one way or another, including presenilins, sphingolipids, ApoE4 and cholesterol. This is not the place to worry about the details – the basic problem here is that the brain is committed to burning glucose via the full forward Krebs cycle (ironically, about the only place in this whole book where the canonical cycle is operating). The brain can't easily switch to some other fuel or flux pattern when neurons or their 'symbiotic' tissues are damaged. The bottom line is that our risk of Alzheimer's disease depends in part on factors such as diet and exercise, which influence our lifelong exposure to high blood glucose. But underpinning this lifestyle risk, we are at the mercy of how effectively our two genomes, mitochondrial and nuclear, deal with the need to transform high glucose into high electrical membrane potential in the pancreas and brain.

---

can be directly toxic to proteins). For a long time, it was thought that insulin does not affect the uptake of glucose by the brain, hence the brain could not be 'insulin resistant', but more recent work indicates that neurons do indeed become insulin resistant, to the point that Alzheimer's disease has been termed 'type 3 diabetes'. One paper by Paula Moreira captures the nub of the problem in her title: 'Sweet mitochondria – a shortcut to Alzheimer's disease'.

Know thyself

The portrait of ageing I have drawn here is fundamentally epigenetic. Ageing is not driven by mutations in genes accumulating over time, but by changes in gene activity – epigenetics. But this word is misleading in two respects. First, 'epigenetics' sounds intrinsically reversible, but plainly that's hardly true with ageing – we might be able to improve our lifestyle and reduce our biological age a little, but only so far. Don't think it's easy to switch epigenetic states. Cells grown in culture can retain their identity as kidney or liver cells for decades, even though they have the same set of genes as neurons. Senescent cells are hard to change too. That relates to my second reservation. The phrase 'epigenetic state' sounds static, but nothing could be further from the truth. A cell can also look static under a microscope, yet its state is the product of more than a billion metabolic reactions every second.[11] You are composed of at least thirty trillion cells, so in the last second your tranquil demeanour was sustained by an incomprehensible one hundred billion trillion reactions (10^{23}, or 100,000,000,000, 000,000,000,000). I'm now in my mid-fifties, so my wrinkles and aches and pains are the product of about 10^{32} reactions to date, roughly a billion times the number of stars in the known universe. I won't even wonder how many of them didn't quite work out properly. It's astonishing I'm still alive at all. This is the metabolic flux that we've been talking about, second

---

11 I take this number from Joana Xavier, one of the stars of a rising generation of scientists who are rethinking the origins of metabolism at the emergence of life, who has recently joined us at UCL. Her estimate of one billion metabolic reactions per second is actually for a bacterial cell, not one of our own more complex eukaryotic cells. Our cells might have as many as 20 billion transformations a second, but the number varies and is less certain so I've stuck with the conservative lower estimate here. It's still extremely big.

by second, day by day, year by year – an astonishingly stable, unceasing flood.

At the very heart of this metabolic whirl is the Krebs cycle, inextricably intwined with the fabric of our mitochondria. Their electrical membrane potential is bound to the flux of reactive oxygen species, and to our capacity for burning 2H to power a hundred billion trillion reactions each second. The Krebs cycle is often not a cycle at all, but a roundabout, a magic roundabout, with flux entering or exiting at every junction, and easily going in either direction. The relative concentrations of Krebs-cycle intermediates, their ratio, is one of the most useful real-time readouts of the steady-state health of a cell. Do not think a steady state is static, any more than the eye of a tornado, the vortex of a whirlpool or the surface of a star is static – it is a stable state produced and maintained by the incessant whirl of activity. It founders and dies when the whirl of activity ends.

In cells, some of these whirling reactions go wrong and damage molecular machinery such as proteins. Not all proteins are replaced effectively, and repairing or replacing them is a drain in itself – one of the most energy-sapping tasks that cells face. Eventually the respiratory machinery itself is damaged, and the rush of electrons to oxygen is impaired. ROS flux creeps up. Cells do what they must and compensate by suppressing respiration a little. NADH is oxidised less effectively and the Krebs cycle loses forward momentum. Intermediates such as succinate start to accumulate and seep out from the mitochondria. They activate proteins such as $HIF_{1\alpha}$, which in turn alter the behaviour of thousands of genes, pushing cells into a senescent state or to their demise. Ageing is hard to reverse for the simple reason that metabolic flux can't stop until it ceases forever. Repair is never perfect.

I mentioned that metabolic rate and lifespan only correlate

loosely, because organisms can vary how much they invest in restricting or repairing damage. Perhaps the most striking examples are bats and birds. Animals with the power of flight live as much as ten times longer than their land-lubber cousins with a similar body size or resting metabolic rate. Recent work from Gustavo Barja in Madrid suggests that these longer-lived animals restrict ROS flux from complex I, simply by down-regulating a subunit containing the most reactive iron–sulfur cluster.[12] Far from suppressing complex I activity, restricting ROS flux enables bats and birds to maintain or increase their aerobic capacity without losing redox tone – their balance of electron sources and sinks. They can go on burning NADH with a Krebs cycle in full forward mode for much longer. That in turn postpones the age-related shift to a senescent epigenetic state. But not surprisingly, there are costs. Restricting ROS flux may have little effect on the accumulation of damaged proteins. Remember that ROS are less damaging than the lurid language used to label them would have you believe. Even so, a faster pace of living entails more turnover of molecular machinery, and therefore more damage. Longer-lived animals must invest more in damage limitation. The cost is ultimately fewer offspring per unit time, and we are back into standard evolutionary biology.

These changes are the work of selection, even if they are

---

12 Barja argues that this is linked to a programme of ageing: that ROS flux from complex I is deliberate. I agree that it is deliberate, but find the evidence for programmed ageing unpersuasive. To my mind, ROS flux signals respiratory capacity relative to demand, facilitating swift changes in respiratory capacity when needed, such as in response to changing oxygen levels or substrate availability. This explanation is based on my colleague John Allen's ideas for why mitochondria retain genes at all; he argues they are needed to control respiration. That would predict there are evolutionary costs in limiting the responsiveness to rapid change, perhaps in terms of adaptability to variable environments or response to infections, but that's another story.

relatively simple and can occur over just a few generations. Our own lifespan is harder to modify, because our physiological limits are set by the genes that we inherited – we can't mix and match over generations to improve our performance. But we can influence metabolic steady states through our behaviour. To do so, we need to know ourselves. Remember that your mitochondria are not the same as mine. We each have distinct mitochondrial DNA, set against kaleidoscopic nuclear genomes. Some mitochondrial DNA is associated with longer life and possibly lower ROS flux. For example, one variant common in Japan is linked with half the risk of age-related diseases and twice the likelihood of surviving to a hundred. In any case, differences in mitochondrial function have inevitable downstream effects on Krebs-cycle flux, gene activity and the rate of ageing. These differences are amplified or dampened by our lifestyle, our diet, activities or habits such as smoking, but what works best for me might be quite different for you. The irony is that in our quest to understand the genes that seem to control ageing, we have binned the very ones that are key to living well – the mitochondrial genes, which channel the unceasing whirl of reactions that keep us alive at all.

Over the entrance to the Temple of Apollo at Delphi, the Greeks inscribed three maxims. The first was 'Know thyself'. They might never have imagined the invisible metabolic flux dancing through genes that defines us as individuals and humans, but they knew well how to live and die. We each need to learn that lesson for ourselves. The second maxim was 'Nothing to excess'. Think about how hard it is to change our epigenetic state: we must sustain a new pattern of flux moment by moment, year after year. Changes that can't be sustained are no use – we will slump back soon enough into the metabolic doldrums of senescence. Excess isn't sustainable over a lifetime.

You will damage yourself, along with your mitochondria. If you want to make yourself more youthful, if not young again; if you want to trim your biological age and shift your epigenetic state to a more aerobic setting; if you want to live well and die well – then you will need to eat well and keep active in your own way. You will need to sustain your lifestyle over decades to invigorate your health with age and add good years to your life.

The third Delphic maxim is a little more ambiguous. It is usually translated as 'Surety brings ruin'. I imagine this to mean that we should never crave certainty, for nothing is certain, least of all in science. Science is not a collection of dusty facts, but a way of exploring the unknown, of making out the contours of a long mysterious coast. I have tried to write this book in that spirit, connecting the first stirrings of life on a geologically restless planet with the glorious pinnacles of evolution, and ultimately our own demise. I can't be right about all of it. But even if the details of the coastline that are emerging through the mist are distorted, this is a thrilling new continent, which transforms the relationship between metabolism and genes: what it is to be alive. We are not islands, but a part of this continent, connected with the main, with all life on our planet from the very beginning. I hope that you'll see yourself a little differently now. And with that in mind, let's end our journey with an unfurling view of the final frontier.

SELF

'I think therefore I am' said Descartes, in one of the most celebrated lines ever written. But what am I, exactly? An artificial intelligence can think too, by definition, and therefore 'is'. Yet few of us could agree whether AI is capable in principle of anything resembling human emotions, of love or hate, fear and joy, of spiritual yearnings for oneness or oblivion, or corporeal pangs of thirst and hunger. The problem is we don't know what emotions are: what is a feeling in physical terms? How does a discharging neuron give rise to a feeling of anything at all? This is the 'hard problem' of consciousness, the seeming duality of mind and matter, the physical makeup of our innermost self. We can understand in principle how an extremely sophisticated parallel processing system could be capable of wondrous feats of intelligence, but we can't answer in principle whether such a supreme intelligence would experience joy or melancholy. What is a quantum of solace?

Why would I even touch on such a question in a book on the Krebs cycle? The answer is that the flux of metabolism, moment by moment, decade after decade, has to correspond in some way to the stream of consciousness – what else could animate our innermost being? In this book, we've explored the dynamic side of biochemistry, the continuous flow of energy and matter that makes us alive. I have argued that this flow began in deep-sea hydrothermal vents, electrochemical flow reactors with structures akin to cells, where the flow of protons across barriers and membranes coaxed H_2 and CO_2 to react to form the Krebs-cycle intermediates at the heart of metabolism across life. These in turn gave rise to amino acids, fatty acids, sugars, nucleotides – the building blocks of life. It might seem uncanny that whole metabolic pathways can spring into existence in this way, in the absence of genes and information, but this is what recent experiments are telling us. There is something thermodynamically and kinetically favoured about the innermost chemistry of life. I find this unsettling, but that's how it is.

The power of this chemistry to self-organise, to grow, to form protocells animated by the same flux of gases into living matter, gave meaning and context to genes and information. For me, the first genes were random strings of a few letters of RNA, polymerising inside protocells growing in those deep-sea vents. From the beginning, genes copied themselves inside protocells, spreading when they promoted cell growth, regenerating what had come before faster and better. Genes never supplanted the deep chemistry of cells. They conserved it, and they built on it. Four billion years on, genes are still faithfully regenerating the deep chemistry of life in our own cells, at an unfathomable rate of billions of reactions per second. From the beginning, the flow of energy and matter through the Krebs cycle was bound to the electrical potential on membranes. Flux is movement. The electrical potential humming away on cell membranes is movement too, dancing charge, electrons and protons, the elementary particles of life. Moving charge generates electromagnetic fields that permeate our being. And clearly, the flux of metabolism generates electromagnetic fields on cells. Could feelings somehow be related to this dance of charge, the ephemeral states of cells?

The idea is pleasing, but I wouldn't have given it any more thought but for a visit from a scientific seer. Luca Turin is a biophysicist interested in quantum biology. He has led an unusual life, working for many years on scent and smell, and the possibility that we may be able to detect quantum vibrational states. When he visited me at UCL, that's what I thought he'd want to talk about. I knew little about it, and did not feel able to comment. But that wasn't what he wanted to talk about at all: he had mitochondria on his mind. I could judge what he had to say about mitochondria, and it was thrilling. Turin is unafraid to confront the unknown in science, or for that matter the 'known' (riling some), but combines his yen for vistas new with a rigorous understanding of some fundamental methods in biophysics, such as electron spin resonance. Rarely have I met someone who thinks in such clear lines. His papers convey this clarity, coupled with wry amusement. 'Almost the only thing we know for sure about consciousness is that it is, so to speak, soluble in ether, chloroform and a variety of other solvents.' Intriguingly, anaesthetics can dissolve consciousness reversibly not only in humans, but even in the simplest animals, as well as single-celled protists such as paramecium. Turin concludes from this that consciousness is not an emergent property of the complex nervous systems of higher animals, but is something more fundamental that works at the level of cells. This means we can study consciousness in some simple experimental models such as fruit flies. As Turin puts it, 'While it is not known to what extent fruit flies are conscious, they are most definitely unconscious when exposed to chloroform or ether.'

The mechanism of general anaesthesia is one of the major unsolved problems in science, ranked as such by the journal

Science alongside cancer, quantum gravity and high-temperature superconductivity. Our skill in manipulation often outstrips understanding, and in this case we know how to control the effects of anaesthetics with exquisite finesse, but next to nothing about how they actually work. The problem is the lack of relationship between molecular structure and biological activity: molecules of disparate sizes and shapes (precluding a common interaction with some receptor in a normal lock-and-key mechanism) all act as general anaesthetics. Perhaps most baffling is the gas xenon. As Turin points out, xenon has no 'shape' (it is a perfect sphere of electron density) and no chemistry – it is an inert gas. But it does have *physics*. It is capable of facilitating electron transfer between conductors. Just think about xenon lamps, which generate a white light similar to sunlight. So in principle xenon could induce anaesthesia by facilitating electron transfer. But why on earth would electron transfer induce anaesthesia?

One of the few things that general anaesthetics do have in common is that they are lipid soluble – they accumulate in membranes, and the strength of anaesthesia depends on their concentration more than their structure. Anaesthetics accumulate in the mitochondrial membranes too. So, could they facilitate electron transfer to oxygen in cell respiration? Turin's work shows that this might be the case. Electron spin resonance gives a signal associated with oxygen, which shifts under anaesthesia (it's the only signal that does), but not in mutant flies resistant to anaesthesia. Even more intriguingly, Turin has detected a radiowave signal associated with electron transfer in respiration. Because all proteins are composed of chiral amino acids (which always have the left-handed form) the transfer of electrons from one protein to another in respiration locks them into the same spin phase, which can be detected as a radiowave signal when they relax upon reacting with oxygen. Don't worry about the details here. The point is that these radiowave signals increase when brain areas are active, and are suppressed by anaesthesia, again implying an effect on respiration. You can see why Turin has a reputation for being difficult. His science is right at the edge of the known. Even he admits that brains emitting radio waves sounds like the stuff of science fiction. But it seems they do.

Let's get back to xenon. All this suggests that xenon concentrates in the hydrophobic pockets of proteins sitting within the mitochondrial membranes, and flits respiratory electrons straight to oxygen. The effect must be subtle, for too much would kill us, and overdose is always a risk with anaesthetics. But suppose it's true. What next? Instead of electron transfer to oxygen being coupled to proton pumping and ATP synthesis, some proportion must hop on a xenon bridge straight to oxygen. That oxygen is presumably still bound to cytochrome oxidase at the end of the respiratory chain in the normal way, so the electrons are not escaping as free radicals. Even so, short-circuiting the respiratory chain must affect the electrical membrane potential, which should be measurable (though these are not easy measurements to make). So ... could it be that a change in mitochondrial membrane potential affects our conscious state?

I mentioned electromagnetic fields. We have long known that the brain generates electrical fields, which we measure in the EEG (electroencephalogram). As with general anaesthetics, we know far more about how to interpret patterns in the EEG associated with epilepsy or sleep than we do about what generates them in the first place. According to the neuroscientist Michael Cohen, we know 'shockingly little about where EEG signals come from and what they mean'. Plainly the EEG is produced by changes in electrical voltage, and these changes are big enough to incriminate large networks of neurons firing in synchrony (rather than individual cells). But these neural networks are still composed of individual neurons, which behave in similar ways. The question is, at the cellular level, which electrical charges are involved? The glib assumption is that charges on the cell membrane (or action potentials) are responsible. But if Turin is right, then a big part of the answer might be mitochondrial membrane potentials. Not only is electron transfer to oxygen implicated in consciousness, but the mitochondrial membrane potential is twice that of neural cell membranes, and the convoluted folds of the mitochondrial inner membrane (the cristae) offer a much larger total surface area of charged membrane.

Moving charge necessarily generates an electromagnetic field, and the mitochondria clearly do so – not only with the transfer of electrons to oxygen but even more dramatically in the circuit of protons across the membrane, looping from the respiratory complexes to the ATP synthase and back round. Doug Wallace is again at the forefront of the field, attempting to measure the strength of electromagnetic fields in individual mitochondria. But there are some broader principles too about the way that electromagnetic fields interact with each other – separate fields can interfere with each other (cancelling out) or can link in phase to generate a stronger field, operating over a longer distance. Parallel membranes, such as the cristae membranes of neuronal mitochondria, should generate stronger fields that amplify a signal and potentially interact with weaker fields on the cell membrane, modulating neural activity. Could this be what generates the EEG? I suspect so, although that would be immaterial if the EEG is no more than an epiphenomenon, a reflection of some underlying activity with no power to influence anything itself. But there's strong evidence that electrical fields can and do play a direct role in brain function. If you cut the axon of a neuron, for example, and separate the two cut ends by a fraction of a millimetre (which is too far for chemicals to cross quickly), an action potential can leap the gap as if it wasn't there. Electrical fields can easily explain this behaviour. If so, the key point is that the electrical fields generated by neurons do have motive force. They are not too weak to change things physically, as long assumed.

This kind of statement might have pushed the boundaries of respectable science until recently, but the extraordinary work of the developmental biologist Michael Levin and others shows that electric fields can control the development of small animals such as the flat worms known as planarians. I suspect that twenty-first-century biology will be the biology of fields. So, let's take it to be possible that the electrical fields generated by mitochondria do have motive force. What can that tell us about consciousness? Well, for a start, it might tell us why the brain is so hooked on glucose as a fuel. If you recall, calcium influx into the mitochondria from their associated membranes (MAMs) activates the enzyme pyruvate dehydrogenase, ramping up Krebs-cycle flux and ATP synthesis nearly exponentially. Plainly that powers work, but it also gives scope to the full dynamic range of mitochondrial membrane potential. To the full range of electrical fields. To the full music of the orchestra. Until now, biology has tended to study the materials that make up the instruments. The time has come to close our eyes and listen to the music. I want to suggest to you that this music is the stuff of feeling, of emotion. Electrical fields are the unifying force that binds the disparate flowing molecules of a cell together to make a self with moods and feelings. Alzheimer's disease is the fading of that music as the fields fragment.

Let's put aside multicellular organisms with their nervous

systems and think about protists such as paramecium, which also generate electrical fields on their mitochondrial membranes. Just watch their amazing behaviour under the microscope, the way they move around, explore, graze, give chase or flee from predators, struggle for their lives, or regenerate themselves, whirring parts and all, after some disastrous encounter. This behaviour is marvellous and sophisticated, and takes place in real time as we watch. What coordinates all of this? Do you think it is coordinated by lock-and-key receptor molecules, genetically specified interactions between proteins? What unifies the whole? What coordinates it as a 'self'? Once you think about electrical fields, it is hard to imagine anything else. But then we are faced with another problem. As with our own nervous systems, I'm arguing that the electrical fields in protists are mostly generated by mitochondria, deep within the cell. So *why* would these internal electrical fields become associated with the self, the life, the potentialities, of an organism as a whole? They're only a bit of a cell, after all. Why would electrical fields in mitochondria, generated by flux through the Krebs cycle, equate to the strivings of the self?

To understand why – why the language of mitochondrial electrical fields became bound to the ephemeral states of cells – you need to know that mitochondria were bacteria once. They were engulfed by other cells nearly two billion years ago. I explored the extraordinary ramifications of that relationship in *The Vital Question*. All we need to know right now is that the electrical potential on mitochondrial membranes inside our own cells is the same as the charge on the plasma membrane of bacteria – the membrane bounding the cell, which separates and links the self that is a bacterium with the outside world. For the bacterial self, the plasma membrane is the threshold of the known universe. Everything else is shadow.

Let me give you an example of how important this membrane potential is to bacteria. In the ocean there are about ten times as many viruses that attack bacteria (phages) as there are bacteria. You might have seen pictures of these viruses attaching themselves in droves to bacteria. Phages are remarkable, mechanical structures, like miniature lunar landers, with the malevolence of H. G. Wells's Martian tripods, anchoring themselves to the surface and injecting their DNA at high pressure into the soft body of the bacterium. Scores of them can line up like an invasion from outer space. The poor bacterial cell doesn't stand much chance, but it does have defences – defences that we have recently learnt to exploit, called CRISPR, which allow for sophisticated gene editing. If the bacterium (or indeed its ancestors) has had an earlier exposure to the virus, it can recognise the viral DNA and marshal a counterattack, chopping up the DNA into bits before the virus can copy itself. But the time window is short. If there are too many phages, then there's only one way out for the bacterium: die, fast, for the good of its kin. How does it do this? It yanks open pores in its cell membrane, collapsing the electrical potential. It dies almost immediately, before the virus has a chance to copy itself and infect its sister cells. As a result of this sacrifice, at least some of its genes may live on in those sister cells – the bacterial equivalent of us laying down our lives for family.

I have long wondered if that collapsing membrane potential 'feels' like something to a bacterium. More than anything else, the humming electrical potential on the membrane betokens the living force. And if it feels like something for a bacterium to die, its living force sucked away, what about other modulations in the electric fields generated by membrane potential? Viruses are not the only things that kill bacteria. They can die by wear and tear too – damage caused by overexposure to bright light, or too little iron for the photosynthetic systems to work properly, or toxins squirted out by neighbouring cells. Each of these can trigger waves of death through marine blooms of cyanobacteria. All operate through much the same mechanisms, collapsing electrical membrane potential to induce death. Presumably, there must also be some 'pre-death' state, where the living processes are tenuous. Beyond that, membrane potential is needed for far more than the basics of ATP synthesis and CO_2 fixation. It powers the bacterial flagellum, allowing cells to move around and seek better conditions, as well as pumping all manner of things in and out of cells, maintaining their homeostasis. Most strikingly, bacteria need their membrane potential to find their own midpoint, to divide in two and generate offspring. Nothing in biology is more sacred than reproduction, and the simplest form of reproduction does not happen without an electrical charge on the membrane. All these states of living and dying are linked with electromagnetic fields. Do they all feel different? How could they not? Metabolism and electromagnetic fields on the membranes bounding cells are intimately entwined and intrinsically meaningful. These are the living states of cells, the stream of consciousness in its most elementary form.

Shrink yourself down to the size of a molecule in the Krebs cycle. Succinate. The cell you're part of is the size of a city, a metropolis on the scale of London, Tokyo or New York. What connects you with another molecule of succinate across the city, twenty miles away? In what sense are you part of the same entity, the same self? You won't even be succinate for long. In a tiny fraction of a second, you will transform into malate, then oxaloacetate, or perhaps an amino acid or sugar. You are a fleeting moment in the kaleidoscope of metabolism, shape-changing a billion times a second. There's no sense in which information binds you. Yet you are still part of an entity, a self.

Your flux through the Krebs cycle is linked, moment by moment, with the balance of metabolism, with how much there is of you compared with the next molecule. You are bound to the flow of electrons, to the pumping of protons, to the electrical charge on the membrane, to the genes switched on and off. Protons scuttle instantaneously around the membrane, equalising the charge at each location, cohering an electromagnetic field that exerts its force everywhere in the cell. The water bound to surfaces inside the cell oscillates in phase, uniting the molecules of metabolism in the symphony. Changes in the outside world – in food, electrons, protons, oxygen, heat or light – all are converted through metabolic flux into dancing electromagnetic fields, shifting the mood, the living states of a bacterium. You have just been part of something magic, the flow of life through a living cell on this restless planet of ours, the rush of change that forges the oneness of self. You are a moment in a life.

ENVOI

'Like Most Revelations'

Richard Howard

It is the movement that incites the form,
discovered as a downward rapture – yes,
it is the movement that delights the form,
sustained by its own velocity. And yet

it is the movement that delays the form
while darkness slows and encumbers; in fact
it is the movement that betrays the form,
baffled in such toils of ease, until

it is the movement that deceives the form,
beguiling our attention – we supposed
it is the movement that achieves the form.
Were we mistaken? What does it matter if

it is the movement that negates the form?
Even though we give (give up) ourselves
to this mortal process of continuing,
it is the movement that creates the form.

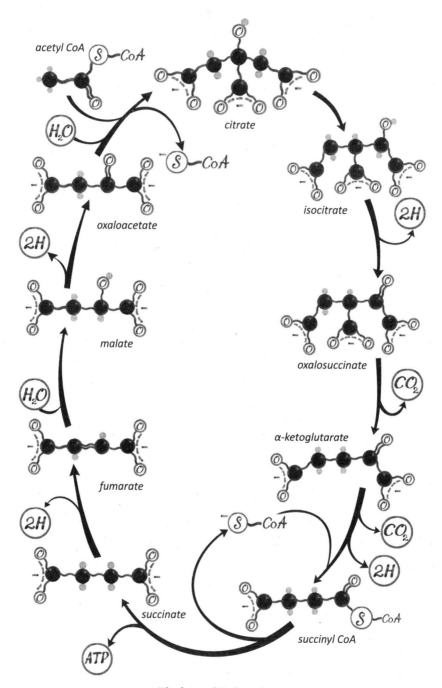

The forward Krebs cycle

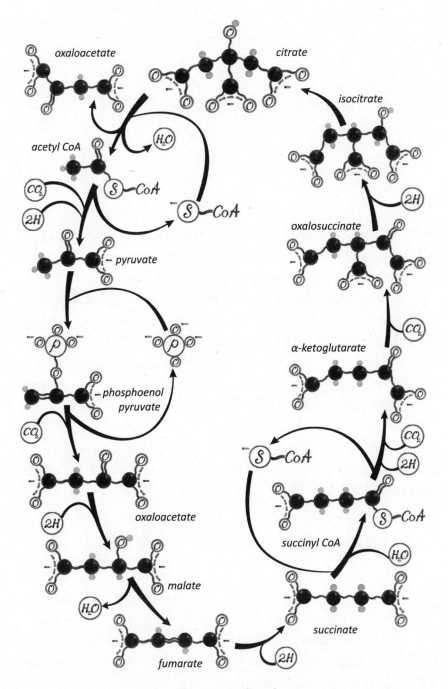

The reverse Krebs cycle

APPENDIX 1

RED PROTEIN MECHANICS

The red protein is ferredoxin, one of the most ancient and fundamental proteins across all of life. It seems to have many jobs, but the common denominator is that it has unparalleled power to transfer electrons onto other molecules, most notably CO_2, to fix carbon in photosynthesis and other forms of autotrophic metabolism. In the main text I gave what might have been a perplexing figure, if you looked at it for a while. Here it is again. Recall that carbon is shown as black balls, hydrogen as grey balls, oxygen as white balls with the symbol 'O'; and R signifies the rest of the molecule, which could be anything you like. Ferredoxin is marked here as Fd, which catalyses the transfer of electrons from the 2H onto CO_2 to extend the chain length of the carbon skeleton:

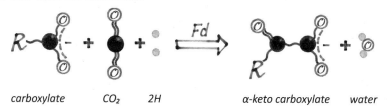

carboxylate   CO₂   2H   α-keto carboxylate   water

What just happened here? You'd struggle to figure it out from the scheme depicted above. In fact, several steps are needed to make this work in modern biochemistry, each one with its own chemical 'prop', notably ATP, coenzyme A (CoA) and ferredoxin itself. These steps are so central to modern metabolism that it's worth considering how each step works.

Let's think about a specific example – how acetate (a C2 molecule) is activated to react with CO_2, to form the C3 pyruvate. In other words, the 'R' depicted in the example above corresponds to a methyl group (-CH_3). The first step goes like this:

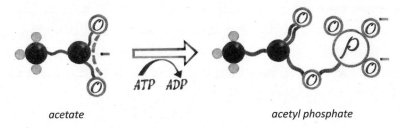

acetate   acetyl phosphate

The problem here is that the carboxylate group (-COO^-) is not very reactive. To react with CO_2, it first needs to be activated. This is achieved initially with ATP, which adds on the phosphate group to form acetyl phosphate, as shown on the right-hand side. (ATP is adenosine triphosphate; the transfer of one phosphate onto acetate by ATP leaves ADP, adenosine diphosphate, which can be recycled to ATP through respiration.)

Having the phosphate attached makes it easier to remove the unreactive carboxylate oxygen and then add on something else in its place. Phosphate is said to be a good leaving group, which is to say, it detaches quite easily as a stable entity, so long as something else is available to take its place. That something else in this case is coenzyme A. I won't show this as a structure as it's quite complex, so I'll mark it simply as CoA-S⁻ to signify that the reactive bit of the molecule is the sulfur (S) atom. In this next step, I'm showing the sulfur atom forming a bridge to coenzyme A:

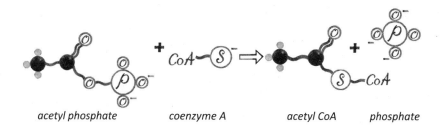

| acetyl phosphate | coenzyme A | acetyl CoA | phosphate |

The important thing to notice here is that the right-hand carbon is now joined directly to a sulfur atom, without there being an oxygen in the way (as there is in acetyl phosphate). And that makes the carbon more willing to react. In the final step, the CoA-S⁻ can now be substituted for a molecule of CO_2, giving the C3 pyruvate:

| acetyl CoA | CO_2 2H | pyruvate | CoA plus protons |

You might still be wondering what the ferredoxin is doing here. It's adding on two electrons, but where did they actually go? The easiest way to see this is to think about what happens when the CoA-S⁻ comes off, and the CO_2 is added on in its place. The reason that CoA-S⁻ can detach relatively easily is that it, too, is a good leaving group: it can pick up both the electrons in its bond with carbon and leave as a stable entity. But that leaves the carbon short of electrons, which it must pick up from somewhere else instantaneously or the reaction won't happen at all:

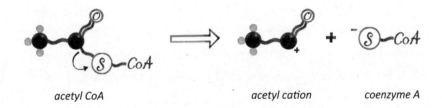

acetyl CoA   acetyl cation   coenzyme A

The curly black arrow shown here denotes the movement of a pair of electrons, in this case the pair of electrons that form the C–S bond. The electrons move wholly onto the sulfur of CoA, giving it a negative charge, and leaving the carbon bereft of an electron pair (hence the positive charge). The reason there's only a single negative charge on the sulfur is that one of the two electrons in the pair belonged to the sulfur in the first place, hence it only receives one additional electron.

Short of an immediate back-reaction (which would just revert to the starting point), where could the positively charged carbon atom pick up the pair of electrons it needs to regain stability? Not from CO_2, because that faces a similar problem. CO_2 is not very reactive, as it has a stable bond structure. But this structure is under some electrical strain. That's because the oxygen atoms have a strong tendency to pull electrons towards themselves, giving them a slightly negative charge (not a full-on electrical charge), which is symbolised by the Greek symbol delta (δ). Each of the two oxygen atoms therefore has a δ^-, whereas the carbon atom is δ^+, as shown below.

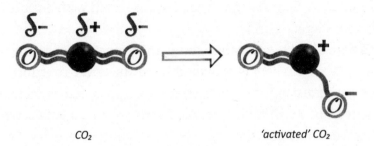

CO_2 'activated' CO_2

This mild electrical polarity can in principle give rise to a more extreme endpoint, where a pair of electrons is pulled wholly onto one of the oxygens, giving the more reactive, unstable structure depicted at the right in the reaction on the previous page.

You can imagine that the positive charges on the acetate shown above and the CO_2 shown here would not go anywhere near each other; but the pair of electrons transferred from ferredoxin makes a new C–C bond possible, joining the two positively charged carbon atoms together, by giving each one a single electron, to form pyruvate:

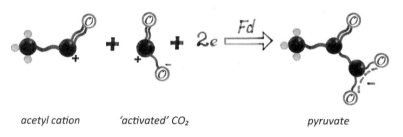

acetyl cation   'activated' CO_2 pyruvate

The chemistry shown here is not intended to be realistic (this is not what actually happens) but to show what the problems are – why this reaction doesn't take place spontaneously and needs to progress through several steps. But something similar may happen in the enzymes, where all the molecular players are held 'just so', meaning that the electrons from ferredoxin are transferred onto the two carbon atoms in the very instant that the CoA-S⁻ grabs the electrons in its bond with acetate and exits the stage.

There's only one other thing to notice here. In the first scheme above, water is shown as a product. Yet that seems to have gone missing in action in the various reaction steps I've depicted above. In fact, water is never produced as a separate entity. Look back at that first scheme – you'll see that the oxygen in the water molecule comes from the carboxylate group – and ends up on the phosphate that detaches from acetyl phosphate. The protons, in that first example, came from the 2H and bind to the O^{2-} to from water. When the oxygen ends up on a phosphate group, those same protons simply balance the charges on the newly released phosphate ion. So the overall reaction is this:

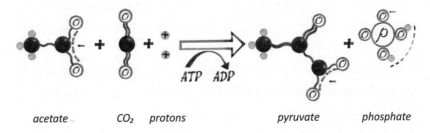

| acetate | CO_2 | protons | | pyruvate | phosphate |

As happens so often in the chemistry of life, the water is there alright, marked by the dashed lines (one of the oxygens plus two of the protons on the phosphate), but it's nowhere to be seen as an independent entity. Biology magics water away.

THE KREBS LINE

In Chapter 3, we followed the fate of CO_2 bound to the surface of a mineral, through to a C2 acetyl group bound to the surface. This acetyl group is equivalent to acetyl CoA, one of the most important molecules in all metabolism, and the end-product of the acetyl CoA pathway. The steps that I drew in Chapter 3 are similar to the acetyl CoA pathway itself, in which a methyl group reacts with CO to form acetyl CoA. In our case, we didn't form acetyl CoA itself (CoA being a complex molecule) but a prebiotic version of the same thing – an activated acetyl group bound to a surface, primed to react.

Acetyl CoA is not only the end-product of the acetyl CoA pathway; it's also a key constituent of the reverse Krebs cycle. Rather than a closed cycle, I talked about a Krebs *line* in Chapter 3, which is to say, a linear pathway that continues the same repetitive chemistry as the acetyl CoA pathway, extending the carbon skeleton from C2 up to at least C4, and possibly even C6. In this appendix, I'll show the first half of the reverse Krebs cycle step by step from an acetyl group bound to the surface of a mineral. I mentioned that the second half of the reverse Krebs cycle repeats the same series of steps as the first half, so I'll leave that to your imagination. Or even better, take out a pen and paper and see how far you get – it's fun!

This really is more of the same. Here's what happens if an

acetyl group is bound on near to another CO (check back to Chapter 3 if you can't recall where the CO comes from). Once again, the acetyl group hops across to join the CO (the same Fischer–Tropsch-type chemistry), this time giving pyruvate (or more technically, a pyruvyl group), again bound onto the surface of the mineral:

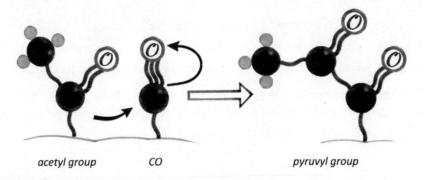

acetyl group   *CO*   *pyruvyl group*

Just in case you don't recognise this as pyruvate, let me draw in the final step again, showing it being released as pyruvate, which I hope you know and love by now:

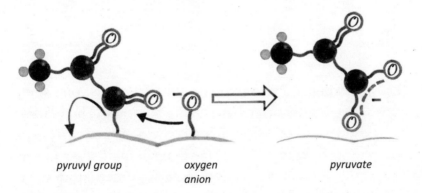

pyruvyl group   *oxygen*   *pyruvate*
                   anion

When bound to the mineral surface, the steps that we've just seen (acetate to pyruvate) are mechanistically equivalent to, but much simpler than, the Krebs cycle itself, which requires ATP, CoA and ferredoxin in succession to force acetate to react. To see why, look at Appendix 1, showing what activation by ATP does. By adding on a phosphate group, ATP allows the removal of an oxygen atom from the acetate, when it is replaced with CoA in the next step. But there's no need for activation by ATP here. When the acetate forms directly on the surface, it comes preactivated, already primed to bind CO_2 and pick up electrons. It's beautifully simple.

The next few steps build on what we've already seen, but alter the pattern slightly, because pyruvate possesses an extra double-bonded oxygen on the first carbon atom next to the carboxylate group, a so-called 'alpha carbon'. This arrangement is unusually reactive. Here's why that Mad-Eye Moody oxygen is so reactive, which brings its own inexorable chemistry.

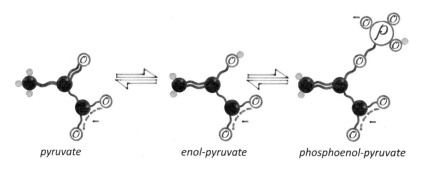

pyruvate   enol-pyruvate   phosphoenol-pyruvate

Let me explain. Pyruvate exists in two states, one being more reactive and shorter lived than the other. The reactive form is known as the 'enol'. This is the activated state seen in phospho*enol*-pyruvate, that branching point in metabolism. In effect, the voracious, electron-hungry oxygen atom drags one of the pairs of electrons in the double bond towards itself. That would normally be prohibited, as the carbon would be left with an unstable positive charge, but in an enol, one proton can flit from the methyl (-CH_3) group onto the oxygen, forming a double bond (the 'ene') between the two carbon atoms and an alcohol (-OH) on the alpha carbon, shown in the middle panel. This structure can be stabilised by phosphate in phospho*enol*-pyruvate, as shown at the right-hand side of the figure opposite.

In the absence of a phosphate, the likelihood of a stable enol form persisting is greater if a neighbouring molecule (be it an enzyme or mineral surface) can abstract a proton from the methyl group, as shown here. In this case, I'm not going to draw the pyruvate with its normal Cheshire Cat smile, but in the more conventional way that emphasises the charge on one of the two oxygens of the carboxylate group. That way you can see it interacting with the surface better:

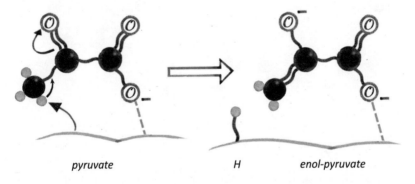

pyruvate   H   enol-pyruvate

The curly arrows here depict a proton acquiring a pair of electrons from the surface; the two electrons that formerly bound the hydrogen to pyruvate now move to form the double bond in enol pyruvate, which is stabilised by the oxygen grabbing an electron, as it likes to do. The proton bound to the surface is now in fact a bound hydride ion (H^-), equivalent to the hydrogen in NADH. The H^- ion can react readily with a proton in the acidic conditions to form hydrogen gas (H_2), which easily bubbles away.

The reactivity of the enol stems from its shape-changing behaviour, specifically the tendency of the carbon–carbon double bond to react with other molecules. In general, carbon–carbon double bonds are not especially reactive: if one of the electron pairs reacted with something else, it would leave a positive charge on the carbon involved. But enols are more reactive precisely because the negatively charged oxygen can immediately reform its double bond with the carbon (which happens less easily if the oxygen picks up a proton to become an alcohol). All that means an enol can bind to a mineral surface, and once bound will be able to react with other bound molecules such as CO, exactly as we've seen before. Here's the enol form of pyruvate binding to a surface and then reacting with CO to form the C4 oxaloacetate:

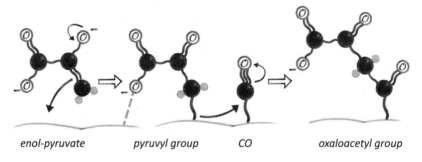

enol-pyruvate   pyruvyl group   CO   oxaloacetyl group

Notice that I'm drawing the carboxylate group with its negative charge in the conventional way again here, to emphasise its electrical interactions with the charged surface. I'll use these styles interchangeably in the next couple of pages. As we saw earlier with pyruvate, oxaloacetate can remain bound to the surface, or it can react with a bound oxygen, being released as the free form of oxaloacetate. I've shown that at the top of the opposite page for completeness.

We're nearly there. We've now synthesised three of the most deeply conserved molecules in all of metabolism (acetate, pyruvate and oxaloacetate) through a repetitive series of reactions that demands little more than CO_2 binding onto a surface that can transfer electrons. The last few steps take us right through

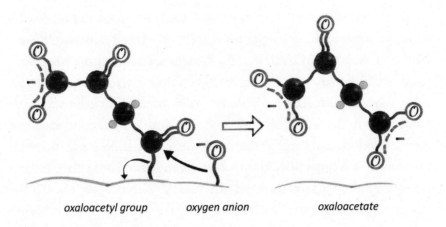

oxaloacetyl group   *oxygen anion*   *oxaloacetate*

to the molecule that started it all, back in the days of Szent-Györgyi and Krebs: succinate. These steps continue to revolve around the double-bonded oxygen. First, another pair of electrons transfers from the surface onto the alpha carbon of oxaloacetate, allowing the oxygen to wrest its electrons from the carbon:

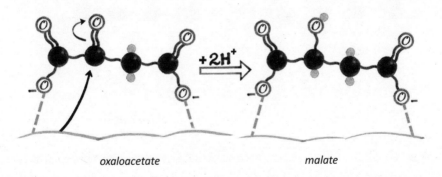

oxaloacetate   *malate*

Remember that we're in a mildly acidic environment, where protons are plentiful. I'm not showing their movements any more, but recall that two electrons plus two protons equates to the addition of two hydrogen atoms, one onto the alpha carbon and the other onto the oxygen, giving the C4 malate (right-hand side).

The final step we'll consider involves the loss of the alcohol (OH) group as a hydroxide ion (OH⁻). As noted before, this is *favoured* in the acidic environment because it can immediately react with a proton to form water, helping to neutralise the acidic conditions. As before, water is formed easily in a watery environment simply through the removal of its component parts. Of course, the OH⁻ can only leave if the electron pair it takes with it is replenished instantly from somewhere else. Here I'm going to link two steps of the reverse Krebs cycle together (skipping neatly over fumarate), in part because I fear your patience with reaction mechanisms may be wearing a little thin, and in part because it's mechanistically simpler for the electrons to transfer directly from the surface to the alpha carbon, to form succinate from malate in one fell swoop, like this:

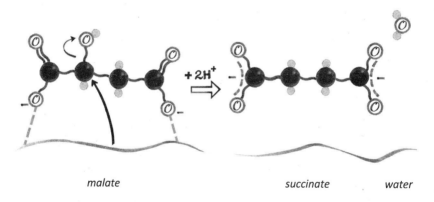

malate   *succinate*   *water*

And that's it! Think for a moment what we've done here. Starting with CO_2 and a common mineral surface (as described in Chapter 3) we've trotted through the first half of the reverse Krebs cycle, forming acetate from scratch, then going on right the way through to succinate. I'm stopping at succinate, for the second half of the cycle repeats the same steps up to aconitate, before shifting slightly for the final step of the full reverse Krebs cycle (see the figure on page 128). I'm not suggesting that prebiotic chemistry will form citrate or the complete reverse Krebs cycle, but the first few steps are exactly as predicted from first principles, based on the known reactivity of carboxylic acids, the known behaviour of mineral surfaces, and at last, the results of some revolutionary experiments.

ABBREVIATIONS

2H Two hydrogen atoms (two electrons plus two protons). Can be extracted from anywhere in a molecule and transferred onto another molecule.

ATP Adenosine triphosphate. The 'universal energy currency'. Has a tail of three phosphate groups linked in a chain. Chopping off one phosphate leaves adenosine diphosphate (**ADP**) plus inorganic phosphate (P_i), which provides the energy needed to drive many reactions in the cell. The ATP synthase regenerates ATP by rejoining a P_i onto ADP.

CoA Coenzyme A. A complex molecule containing a reactive thiol (**-SH**) that can bind to acetate (vinegar) to form acetyl CoA, one of the most important molecules in all of metabolism.

CO₂ Carbon dioxide. The 'Lego brick' from which almost all organic molecules are made in biochemistry via photosynthesis or by reacting with gases such as H_2 in hydrothermal vents.

-COOH Carboxylic acid group. Characteristic of all Krebs cycle molecules, which can have two or three such groups. The proton (H^+) dissociates easily, giving the carboxylate (**-COO⁻**) group.

DNA Deoxyribonucleic acid. The hereditary material. Forms the famous double helix, composed of long chains of millions of letters (nucleotides) that pair together to form complementary strands. When prised apart, each strand acts as a template, allowing the precise sequence to be copied.

Ech Energy-converting hydrogenase. A membrane protein that uses the flow of protons (H^+) to power the transfer of electrons from H_2 to ferredoxin, which in turn passes them on to CO_2. A prebiotic version could have driven CO_2 fixation at the origin of life.

Fd Ferredoxin. The red protein. An iron–sulfur protein with a unique power to transfer electrons onto CO_2 to form organic molecules. The form that transfers electrons is shortened to Fd^{2-}.

FeS cluster One of several different types of small inorganic clusters of iron and sulfur, containing just a few atoms, with structures similar to minerals. The classic is the **4Fe4S** cluster.

GSH Glutathione. GSH is the reduced form (containing H) and **GSSG** is the oxidised form (where an H has been extracted from each of two GSH molecules). An important cellular antioxidant.

GWAS Genome-wide association studies. Statistical studies that correlate single-letter differences in DNA across whole genomes with the risk of diseases.

H^+ Proton; a positively charged particle found in the nuclei of all atoms. The nucleus of the hydrogen atom is composed of a single proton.

H_2 Hydrogen gas, composed of two hydrogen atoms covalently bound together.

MAM Mitochondria-associated membranes. Part of an intracellular membrane system called the endoplasmic reticulum, where it is closely appositioned with the mitochondria.

NAC N-acetyl cysteine. An antioxidant that can be taken as a dietary supplement. Too much is toxic.

NADH Nicotinamide adenine dinucleotide. One of the most important carriers of 2H, NADH is the form loaded with 2H, whereas NAD^+ has passed on its 2H (becoming oxidised). NAD^+ picks up 2H from Krebs-cycle intermediates and then passes on the electrons to complex I in respiration. Technically, I should say that NADH does not carry 2H, but rather two electrons and one proton, known as a 'hydride' ion; but because there is almost always a proton nearby, the transfer of a hydride ion plus the nearby proton equates to the transfer of 2H.

NADPH Nicotinamide adenine dinucleotide phosphate. A 2H carrier used in biosynthesis. **NADP⁺** is the oxidised form (lacking 2H). The ratio of NADPH to NADP⁺ is pushed far from equilibrium, giving NADPH more power to force its 2H onto other molecules than NADH. Note that, as with NADH, NADPH actually transfers a hydride ion (two electrons plus one proton), but because there is nearly always a proton nearby, it amounts to the transfer of 2H.

pH A log scale for measuring the concentration of protons in water. Water (H_2O) splits into H^+ and OH^- ions. At neutral pH (7) their concentration is equal. At acidic pH (< 7) there is an excess of H^+ ions; at alkaline pH (> 7) there are more OH^- ions.

RNA Ribonucleic acid. A working copy of a gene, composed of a single strand of nucleotides linked together in a chain. Can fold into complex shapes and catalyse reactions (as a ribozyme) as well as acting as a hereditary molecule (in some viruses and in a hypothetical early 'RNA world').

ROS Reactive oxygen species. Reactive forms of oxygen that mostly form when oxygen grabs single electrons from FeS clusters in the respiratory chain to form 'free radicals' and related molecules.

-SH A thiol, containing a sulfur joined to a hydrogen, which is quite reactive. Found in coenzyme A, the amino acid cysteine (and so many proteins) and the antioxidant glutathione.

SNP Single-nucleotide polymorphism (pronounced 'snip'). A single-letter change in DNA, which can differ between individuals. We have millions of SNPs, making us genetically different to each other.

FURTHER READING

This is not a textbook or research paper, so is not referenced in a comprehensive way. But if you've read this far, you will have become deeply embroiled in the wider literature. This literature can be overwhelming. It is hard to know where to start, or even how to access it. I've therefore given a list of the papers and books that have most influenced me. Given that this is a personal response to the literature, it felt only right that I should add a few brief words to each of my selections below. I hope these notes will prompt you to read a few pieces that you might otherwise not have given a second glance. I've organised them by themes as you encounter them in each chapter.

Introduction: Life itself

Leeuwenhoek and the inscrutability of cells

Clifford Dobell, *Antony van Leeuwenhoek and his 'Little Animals'* (New York, Russell & Russell, 1958). A delightful if antiquated biography of van Leeuwenhoek by a distinguished protistologist who learnt Dutch to read Leeuwenhoek's letters to the Royal Society. It was first published in 1932 on the 300th anniversary of Leeuwenhoek's birth. Dobell was perhaps the first to identify exactly which protists van Leeuwenhoek had seen. A work of love. My edition from 1958 also features an Introduction by Cornelis van Niel, one of the pioneers of photosynthesis research in bacteria.

Brian J. Ford, *Single Lens: The Story of the Simple Microscope* (New York, Harper & Row, 1985). A remarkable story by a master of microscopy, in which Ford rediscovers nine of Leeuwenhoek's original samples hidden in the library of the Royal Society. He uses single-lens microscopes designed by Hooke and Leeuwenhoek to establish exactly what they had been able to see, while setting the story in its rich historical context.

N. Lane, 'The unseen world: reflections on Leeuwenhoek (1677) "Concerning little animals"', *Philosophical Transactions of the Royal Society B* 370 (2015), 20140344. My own attempt to assess what van Leeuwenhoek's discoveries mean for biologists today, written for the 350th anniversary edition of the *Philosophical Transactions of the Royal Society*, the world's oldest continuously published scientific journal. Includes a light-touch historical overview.

The paradox of information in biology

Paul Davies, *The Demon in the Machine: How Hidden Webs of Information are Finally Solving the Mystery of Life* (London, Penguin, 2019). An eloquent manifesto for reconceiving the role that information plays in life and its origin, by a leading physicist and thinker who has long considered the makeup of life. Not surprisingly, I don't agree with all he writes, but I notice we are both searching for meaning in fields, whether they be electrical or informational (or the same thing).

Erwin Schrödinger, *What Is Life?* (Cambridge, Cambridge University Press, 1967). An ageless classic; one of the most influential science books of the twentieth century, which still rewards reading today. Wrong on plenty of details, but an unparalleled example of how far vision and clear thinking can take you in science. His logic reminds me of Lucretius (*On the Nature of Things*).

L. D. Hansen, R. S. Criddle and E. H. Battley, 'Biological calorimetry and the thermodynamics of the origination and evolution of life', *Pure and Applied Chemistry* 81 (2009) 1843–1855. A technical paper that disagrees with Schrödinger's take on the entropy of living systems – it is not nearly as low as you might think. *Living* costs energy, whereas maintaining the organised structure of cell components such as membranes and proteins has negligible net cost.

Further reading

The dynamic side

Hopkins & Biochemistry 1861–1947. Papers concerning Sir Frederick Gowland Hopkins, OM, PRS, with a selection of his addresses and a bibliography of his publications. (Cambridge, W. Heffer & Sons, 1949). The title says it all. Many of Gowland Hopkins's addresses and papers obsess about the dynamic side. This volume also captures the love and esteem in which he was held.

The unity of biochemistry

D. D. Woods, 'Albert Jan Kluyver', *Biographical Memoirs of Fellows of the Royal Society* 3 (1957) 109–128. A wonderful resource from the Royal Society – short biographical obituaries of their fellows since 1932, freely available here: https://royalsocietypublishing.org/journal/rsbm. This one is on the pioneer of the biochemistry of microorganisms, Albert Kluyver. His original paper on unity was in Dutch so I'm not listing it here.

H. C. Friedmann, 'From "butyribacterium" to "E. coli": an essay on unity in biochemistry', *Perspectives in Biology and Medicine*, 47 (2004) 47–66. An entertaining history of a slippery idea.

R. Y. Stanier and C. B. van Niel, 'The concept of a bacterium', *Archiv für Mikrobiologie* 42 (1962) 17–35. A paper of wonderful insight, which to my mind brought to an end the classical era of microbiology, ceding precedence to the power of genetic information. Ironically, the scrambling power of lateral gene transfer has refocused attention on physiology as an organising principle in bacterial evolution.

The digital jungle

Horace Freeland Judson, *The Eighth Day of Creation. Makers of the Revolution in Biology*, 25th Anniversary Edition (Cold Spring Harbor Laboratory Press, 1996). A magisterial work, which through extensive conversations with the protagonists, deft storytelling, a vast canvas and shrewd judgement conveys a strong sense of the scientific excitement of the dawn of molecular biology.

Matthew Cobb, *Life's Greatest Secret. The Race to Crack the Genetic Code* (London, Profile Books, 2015). A fascinating and authoritative history, which despite being further removed from the events than Judson, combines excitement with the benefit of (wise) hindsight, while bringing the genetics right up to date.

F. H. C. Crick, J. S. Griffith and L. E. Orgel, 'Codes without commas', *Proceedings of the National Academy of Sciences USA* 43 (1957) 416–427. One of the most brilliantly wrong papers in the history of science, which explained how a triplet code that could encompass 64 amino acids actually coded the 'magic number' of only 20. Crick later wrote 'It was a beautiful idea which was completely wrong!'

Molecular machines

Venki Ramakrishnan, *Gene Machine. The Race to Decipher the Secrets of the Ribosome* (London, OneWorld, 2018). A gripping story of the race to resolve the structure of life's protein-building factories, ending in the Nobel Prize for Ramakrishnan. An exhilaratingly honest and thoughtful account, at times reminiscent of Watson's *The Double Helix*.

David S. Goodsell, *The Machinery of Life* (New York, Copernicus Books, 2009). Goodsell is a wonderful artist, whose utterly precise images bring the molecular machinery of life alive in a unique and instantly recognizable way; he even writes well. Is there a biochemist unaware of his work?

The molecular genetic paradigm of medicine

David Weatherall, *Science and the Quiet Art. Medical Research and Patient Care* (Oxford, Oxford University Press, 1995). An eloquent and thoughtful introduction to the molecular genetic paradigm that underpins modern medicine.

James D. Watson, Tania A. Baker, Stephen P. Bell, Alexander Gann, Michael Levine and Richard Losick, *Molecular Biology of the Gene*, 7th edn (Pearson, 2013). The classic molecular biology textbook, originally by Watson and now a larger group; as close to the paradigm of molecular medicine as it possible to get (and justly famous).

Deep unity of biochemistry – the conserved city-centre plan

E. Smith and H. J. Morowitz, 'Universality in intermediary metabolism', *Proceedings of the National Academy of Sciences USA* 101 (2004) 13168–13173. Smith and Morowitz take the case for the unity of biochemistry right back to the origin of life.

W. Martin and M. J. Russell, 'On the origin of biochemistry at an alkaline hydrothermal vent', *Philosophical Transactions of the Royal Society B* 362 (2007) 1887–1925. A monumental paper that has probably influenced my own thinking more than any other. Manages to be thrillingly original while remaining always balanced and scholarly. A masterpiece.

The Krebs cycle

Steven Rose, *The Chemistry of Life*, new edn (London, Penguin, 1999). A classic introduction to biochemistry that caught my imagination when I was at school, and to which I pay oblique homage in the subtitle of this book.

Philip Ball, *Molecules: A Very Short Introduction* (Oxford, Oxford University Press, 2003). Originally published as *Stories of the Invisible*, Ball covers the Krebs cycle and some basic metabolic biochemistry in brief as part of a wider canvas. Ball just doesn't do small canvases.

Harold Baum, *The Biochemists' Songbook*, 2nd edn (London Taylor and Francis, 1995). A classic, witty setting of intermediary biochemistry to music, through which generations of biochemistry students desperately crammed for exams. Just to give you a feel, the Krebs cycle is set to *Waltzing Matilda* and begins: 'Once a jolly pyruvate enters the matrix / Of a mitochondrion, so they say / A decarboxylating, complex dehydrogenase / converts it to acetyl co-enzyme A.' Appropriately enough, my edition comes with a forward from Sir Hans Krebs himself.

Konrad Bloch, *Blondes in Venetian Paintings, the Nine-Banded Armadillo and Other Essays in Biochemistry* (New Haven, CT, Yale University Press, 1997). Seriously! An enjoyable and informative read from a distinguished biochemist, who makes it very clear why most biochemical pathways don't run in both directions.

If pigs could fly chemistry

L. Orgel, 'The implausibility of metabolic cycles on the prebiotic Earth', *PLoS Biology* 6 (2008) e18. Orgel didn't mince his words, also describing the idea that metal ions or minerals could catalyse a whole suite of prebiotic reactions as 'an appeal to magic'. He died in 2007, just before this final paper was published, which, in the way of science, has since been proved at least partly wrong.

Chapter 1: Discovering the nanocosm

Hopkins's biochemistry lab

F. Gowland Hopkins, 'Atomic physics and vital activities', *Nature* 130 (1932) 869–871. The text of Gowland Hopkins's anniversary address to the Royal Society in 1932.

H. H. Dale, 'Frederick Gowland Hopkins 1861 –1947', *Obituary Notices Fellows Royal Society* 6 (1948) 115–145. An affectionate biography from Sir Henry Dale, another pioneering biochemist who was instrumental in defending the freedom of science in the post-war years. If you ever wonder about the value of scientific freedom from political interference, I urge you to read his Pilgrim Trust Lecture to the US National Academy of Sciences, published in the *Proceedings of the American Philosophical Society* 91 (1947) 64–72, in the aftermath of the bomb.

Hans Krebs, *Reminiscences and Reflections* (Oxford, Clarendon Press, 1981). An enjoyable and interesting read, published just after Krebs died, and perhaps for that reason containing a number of uncharacteristic errors in the equations.

Soňa Štrbáňová, *Holding Hands with Bacteria. The Life and Work of Marjory Stephenson* (Berlin, Springer, 2016). A really worthwhile biography of one of the most brilliant and influential female scientists in Britain in the first half of the twentieth century. (But why is it so expensive?)

Krebs in Cambridge

H. Blaschko, 'Hans Krebs: nineteen nineteen and after', *FEBS Letters* 117 (suppl) (1980) K11–K15. Very charming recollections of the young Hans Krebs from a lifelong friend in both Germany and Britain.

H. Kornberg and D. H. Williamson, 'Hans Adolf Krebs, 25 August 1900–22 November 1981', *Biographical Memoirs of Fellows of the Royal Society* 30 (1984) 349–385. Concludes with an appraisal of Krebs's approach to science, asking how he attracted the affection and lifelong loyalty of so many people. From personal experience, they write 'It was in a sense reassuring, and endearing, to discover that Hans was not an accomplished experimenter or even a very practical person. ... Above all, there shone through all his writings and actions a burning passion for truth, a distrust and dislike of all that was pompous and spurious, and a transparent goodness that made Hans a father-figure regarded not only with respect but with affection by those who had long left his group as well as by those who remained.' What a beautiful accolade.

Otto Warburg and respirometry

Otto Warburg, The Oxygen-Transferring Ferment of Respiration, Nobel Lecture, 10 December 1931. Available from: www.nobelprize. org/prizes/medicine/1931/warburg/lecture/. A marvellous and inspiring piece, which captures all that is best about Warburg.

P. Oesper, 'The history of the Warburg apparatus. Some reminiscences on its use', *Journal of Chemical Education* 41 (1964) 294–296. An enjoyably breezy short history that explains how Warburg's name came to be associated with an instrument invented by T. G. Brodie and others.

H. Krebs, 'Otto Heinrich Warburg. 1883–1970', *Biographical Memoirs Fellows of the Royal Society* 18 (1972) 628–699. Krebs's generous tribute to his scientific mentor; even Krebs was unable to present Warburg in a wholly positive light, though he tried hard.

Szent-Györgyi story

Albert Szent-Györgyi, 'Lost in the Twentieth Century', *Annual Review of Biochemistry* 32 (1963) 1–15. A short autobiography of one of the most brilliant and restless characters in biochemistry, including an account of his spying exploits in the second world war.

Albert Szent-Györgyi, *The Living State. With Observations on Cancer* (London and New York, Academic Press, 1972). Some mesmerising lines but I have to say a bit of an eccentric book. He thought about life in the right way, but to my mind never took Peter Mitchell's thinking about electrical membrane potential on board.

Albert Szent-Györgyi, *The Crazy Ape* (New York, Philosophical Library, 1970). Quite a disturbed book with short, revolutionary essays.

Albert Szent-Györgyi, Oxidation, Energy Transfer, and Vitamins, Nobel Lecture, 11 December 1937. Available from: www.nobelprize. org/uploads/2018/06/szent-gyorgyi-lecture.pdf. Szent-Györgyi won the Nobel prize in 1937, the same year that Krebs published his cycle. Here Szent-Györgyi ponders how it all fits together with his own earlier work.

J. J. Farmer, B. R. Davis, W. B. Cherry, D. J. Brenner, V. R. Dowell and A. Balows, '50 years ago: the theory of Szent-Györgyi', *Trends in Biochemical Sciences* 10 (1985) 35–38. A short but dense dip into biochemical history. Puts Szent-Györgyi's conception of carboxylic acids being the carriers of 2H to oxygen into context.

Discovery of the Krebs cycle

H. A. Krebs and W. A. Johnson, 'The role of citric acid in intermediary metabolism in animal tissues', *Enzymologica* 4 (1937) 148

−156. The classic non-*Nature* paper on the Krebs cycle, written with
W. A. Johnson, who did most of the experimental work. Johnson's
contribution was perhaps not fully recognised, as pointed out by
Milton Wainwright in *Trends in Biochemical Sciences* 18 (1993)
61–62, 'William Arthur Johnson – a postgraduate's contribution to
the Krebs cycle.'

H. A. Krebs, 'The intermediate metabolism of carbohydrates',
Lancet 230 (1937) 736–738. An immediate perspective on the cycle,
written by Krebs alone and aimed at medics.

Frederic L. Holmes, *Hans Krebs: The Formation of a Scientific Life
1900–1933*. Volume 1 (Oxford, Oxford University Press, 1991). A
monumental tome of extraordinary scholarship, drawing on Krebs's
lab books in detail, with all their exclamation marks. This volume
deals with Krebs's earlier years, leading up to the urea cycle.

Frederic L. Holmes, *Hans Krebs: Architect of Intermediary Metabolism 1933–1937*. Volume 2 (Oxford, Oxford University Press, 1993).
The second monumental volume of Holmes's scientific biography,
covering the Krebs cycle itself, more or less experiment by experiment. Hard work, but captures a lot of the excitement of science.

Hans A. Krebs, The Citric Acid Cycle, Nobel Lecture, 11 December 1953. Available from: https://www.nobelprize.org/prizes/
medicine/1953/krebs/lecture/. A nice history, including thoughts on
the role of the cycle in biosynthesis as well as the evolution of energy
metabolism. He finishes with the prescient lines: 'The presence of the
same mechanism of energy production in all forms of life suggests
two other inferences, firstly, that the mechanism of energy production has arisen very early in the evolutionary process, and secondly,
that life, in its present forms, has arisen only once.'

H. A. Krebs, 'The history of the tricarboxylic acid cycle', *Perspectives
in Biology and Medicine* 14 (1970) 154–170. Some standard history,
but also unusually personal reflections on philosophical outlook and
motivation, comparing his own approach with James Watson's, in
The Double Helix, published two years earlier.

J. M. Buchanan, 'Biochemistry during the life and times of Hans Krebs and Fritz Lipmann', *Journal of Biological Chemistry* 277 (2002) 33531–33536. Some interesting recollections on these two pioneers of intermediary metabolism from another distinguished biochemist, John Buchanan, who was an early pioneer of carbon isotopes in the study of purine biosynthesis (needed for making ATP).

Fritz Lipmann and acetyl CoA

G. D. Novelli and F. Lipmann, 'The catalytic function of coenzyme A in citric acid synthesis', *Journal of Biological Chemistry* 182 (1950) 213–228. The paper that finally made sense of the first step of the Krebs cycle.

Fritz Lipmann, *Wanderings of a Biochemist* (New York, John Wiley and Sons, 1971). Lipmann's autobiography mixed with essays and papers. An eclectic read, full of interest both historical and scientific. Includes his paper on the origin of life, which begins with 'My basic motivation for entering into this discussion is an uneasy feeling about the apparent tenet that a genetic information transfer system is essential at the very start of life.' Plus ça change.

W. P. Jenks and R. V. Wolfenden, 'Fritz Albert Lipmann', *Biographical Memoirs* 88 (2006) 246–266. The US National Academy of Sciences equivalent to the *Royal Society Memoirs*. Features the memorable closing line that Lipmann died 'at the age of eighty-seven, not long after having learned that his latest grant application had been successful'. Here at least, times have changed.

Peter Mitchell and chemiosmotic coupling

Peter Mitchell, 'David Keilin's respiratory chain concept and its chemiosmotic consequences', Nobel Lecture, 8 December 1978. Available here: https://www.nobelprize.org/uploads/2018/06/mitchell-lecture.pdf. The fascinating story of Mitchell's own thinking, paying homage to the vision of his mentor David Keilin. Also interesting is that Mitchell was clearly thinking about conformational changes in membrane proteins in the 1950s, but later erroneously rejected his own early thinking, living up to his closing line, 'The obscure we see eventually, the completely apparent takes longer.'

Peter Mitchell, *Chemiosmotic coupling in Oxidative and Photo-synthetic Phosphorylation* (Glynn Research Ltd, 1966). The first of Mitchell's two highly influential 'little grey books', published privately in 1966 and 1968. A detailed exposition of redox loops and electrochemistry, but mostly incorrect on the actual mechanisms of pumping.

John Prebble and Bruce Webber, *Wandering in the Gardens of the Mind* (Oxford, Oxford University Press, 2003). A fascinating, detailed biography of Mitchell, scholarly, sympathetic, and extremely knowledgeable, from two biochemists steeped in membrane bioenergetics and its history. They have also published a number of probing articles on the history of bioenergetics, often grounded in the correspondence between the characters involved, all of which are worth reading.

R. E. Davies and H. A. Krebs, 'Biochemical aspects of the transport of ions by nervous tissue', *Proceedings of the Biochemical Society* 50 (1952) xxi. An abstract that captures the essence of the chemiosmotic idea nearly a decade before Mitchell; but I'm not surprised Mitchell missed it. This is a stealth abstract.

A. T. Jagendorf, 'Chance, luck and photosynthesis research: an inside story', *Photosynthesis Research* 57 (1998) 215–229. An entertaining personal account of bioenergetics around the time of Mitchell's hypothesis, and Jagendorf's own seminal experiment that first persuaded a sceptical field. Jagendorf is enjoyably self-deprecating and caustic in equal measure.

Mitchell and Moyle

P. D. Mitchell and J. Moyle, 'Stoichiometry of proton translocation through the respiratory chain and adenosine triphosphatase systems of rat liver mitochondria', *Nature* 208 (1965) 147–151. One of the classic papers published by Mitchell and Moyle in *Nature* in the mid-1960s, less heavily cited than Mitchell's original paper on chemiosmotic coupling or his 'little grey books', but to my mind Moyle's vital experimental contribution has not been sufficiently recognised.

P. D. Mitchell and J. Moyle, 'Evidence discriminating between the chemical and the chemiosmotic mechanisms of electron transport phosphorylation', *Nature* 208 (1965) 1205–1206. A shorter paper, again presenting experimental evidence for the chemiosmotic hypothesis.

P. D. Mitchell and J. Moyle, 'Chemiosmotic hypothesis of oxidative phosphorylation', *Nature* 213 (1967) 137–139. A detailed rebuttal of criticisms, containing a summary of Moyle's experimental evidence.

Paul Boyer and the ATP synthase

Paul D. Boyer, Energy, Life and ATP, Nobel Lecture, 8 December 1997. Available here: https://www.nobelprize.org/uploads/2018/06/boyer-lecture.pdf. A remarkable story of how Boyer conceived the mechanism of rotational catalysis of the ATP synthase, including his early drawings. Sir John Walker shared the prize that year for his X-ray crystallography structure showing that Boyer's conception was right.

J. N. Prebble, 'Contrasting approaches to a biological problem: Paul Boyer, Peter Mitchell and the mechanism of the ATP synthase', *Journal of the History of Biology* 46 (2013) 699–737. Another fine piece on the history of bioenergetics from John Prebble.

The mechanism of proton pumping

M. Wikström, 'Recollections. How I became a biochemist', *IUBMB Life* 60 (2008) 414–417. A memoir of the later 'ox phos wars', over the mechanism of proton pumping, especially cytochrome oxidase. Mitchell resisted the idea of conformational change in proteins for eight years, before eventually conceding that Wikström and colleagues' interpretations were correct. Mitchell viewed the relative positions of electron donors and acceptors ranged across the membrane (redox loops) as central to the chemiosmotic hypothesis, when in fact they are not; the idea stands anyway.

Further reading

Franklin M. Harold, *The Vital Force: A Study of Bioenergetics* (New York, W. H. Freeman, 1986). An unusually personable textbook from one of the pioneers of membrane bioenergetics, which gives a clear view of the chemiosmotic hypothesis. Harold has written several books for the general public which are eloquent, thoughtful and nuanced attempts to confront the biggest questions in biology, albeit with a whimsical tendency to prefer questions to answers. His latest is *On Life: Cells, Genes and the Evolution of Complexity* (Oxford University Press, 2021), a unique distillation from a remarkable nonagenarian.

Evolution of the Krebs cycle

J. E. Baldwin and H. A. Krebs, 'The evolution of metabolic cycles', *Nature* 291 (1981) 381–382. An influential paper that stands out in my mind for not even mentioning the reverse Krebs cycle – which had been proposed fifteen years earlier. How blind are even the finest scientists.

H. A. Krebs and H. L. Kornberg, *Energy Transformations in Living Matter: A Survey* (Berlin, Springer, 1957). Rather surprisingly described by James Watson as the only classic text in biochemistry, this is a detailed molecular exploration, which benefits from Hans Kornberg's microbiological perspective.

Endosymbiosis and mitochondria

Nick Lane, *The Vital Question: Why Is Life the Way it Is?* (London, Profile Books, 2015). Published by W. W. Norton in the US with the subtitle *Energy, Evolution and the Origins of Complex Life*. My own take on how bioenergetics shaped the origin of life and the evolution of complexity.

Nick Lane, *Power, Sex, Suicide: Mitochondria and the Meaning of Life* (Oxford, Oxford University Press, 2018). Reprinted with a new introduction as part of the Oxford Landmark Science series (quite an honour). An older take on the importance of mitochondria, but with more background on both Mitchell and Margulis.

Chapter 2: The path of carbon

Rubisco

S. G. Wildman, 'Along the trail from Fraction I protein to Rubisco (ribulose bisphosphate carboxylase-oxygenase', *Photosynthesis Research* 73 (2002) 243–250. Or how rubisco got its name. A fascinating historical piece in *Photosynthesis Research*, which under the passionate auspices of Govindjee surely has a better and more human record of its own history than any other field.

R. J. Ellis, 'The most abundant protein in the world', *Trends in Biochemical Sciences* 4 (1979) 241–244. A delightfully precise piece that explores why rubisco is so inefficient. John Ellis is one of the clearest thinkers I know, and I strongly recommend his short book on evolution, *How Science works: The Nature of Science and the Science of Nature* (Springer, 2016), which is especially good on why Occam's razor (always seek the simplest explanation) is so central to scientific thinking.

R. J. Ellis, 'Tackling unintelligent design', *Nature* 463 (2010) 164 –165. An interesting commentary from John Ellis on attempts to improve the activity of Rubisco by refolding the cyanobacterial version using chaperone proteins – discovered by Ellis himself.

Lawrence and the Rad Lab

Martin D. Kamen, *Radiant Science, Dark Politics. A Memoir of the Nuclear Age* (Berkeley, University of California Press, 1985). A remarkable and absorbing scientific biography, with many memorable character portraits of the leading physicists and chemists of the day, set in times of heightened politics and war. A unique record.

Oliver Morton, *Eating the Sun: How Plants Power the Planet* (London, Fourth Estate, 2009). A delightful book, as gripping as a novel on the history of photosynthesis, before broadening the sweep to a planetary scale with a compelling analysis of how to counter the climate emergency.

Angela N. H. Creager, *Life Atomic: A History of Radioisotopes in*

Science and Medicine (Chicago, University of Chicago Press, 2015). A scholarly history of the development of radioisotopes in relation to the emergence of Cold War science and the 'porous civilian–military divide'.

Early carbon isotope work

S. Ruben, W. Z. Hassid and M. D. Kamen, 'Radioactive carbon in the study of photosynthesis', *Journal of the American Chemical Society* 61 (1939) 661–663. The first publication from Sam Ruben, Zev Hassid and Martin Kamen on the (crazy) use of ^{11}C for tracking the path of carbon in photosynthesis.

C. B. Van Niel, S. Ruben, S. F . Carson, M. D. Kamen and J. W. Foster, 'Radioactive carbon as an indicator of carbon dioxide utilization: VIII. The role of carbon dioxide in cellular metabolism', *Proceedings of the National Academy of Sciences USA* 28 (1942) 8–15. A thoughtful exploration of how CO_2 might be incorporated into organic molecules based on early radioisotope work ... comes tantalisingly close to proposing a reverse Krebs cycle.

S. Ruben, 'Photosynthesis and phosphorylation', *Journal of the American Chemical Society* 65 (1943) 279–282. Sam Ruben gets so far as to suggest that the molecule that accepts CO_2 in photosynthesis could be a sugar phosphate.

S. Ruben and M. D Kamen, 'Long-lived radioactive carbon: C14', *Physical Review* 59 (1941) 349–354. The paper that first introduced ^{14}C to the world.

The path of carbon in photosynthesis

J. A. Bassham, A. A. Benson, L. D. Kay, A. Z. Harris, A. T. Wilson and M. Calvin, 'The path of carbon in photosynthesis. XXI. The cyclic regeneration of carbon dioxide acceptor', *Journal of the American Chemical Society* 76 (1954) 1760–1770. Number 21! Finally they nailed it. It should be said this is not slow for a major breakthrough. The road less travelled is easily lost. I can only hope that those who fund research never delude themselves into thinking otherwise.

A. A. Benson, 'Following the path of carbon in photosynthesis: a personal story', *Photosynthesis Research* 73 (2002) 29–49. Benson said little about his time with Calvin until decades later, when he finally broke his silence. One senses he was trying hard not to be bitter.

A. A. Benson, 'Last days in the old radiation laboratory (ORL), Berkeley, California, 1954', *Photosynthesis Research* 105 (2010) 209–212. More fascinating reminiscences from Andrew Benson.

T. D. Sharkey, 'Discovery of the canonical Calvin–Benson cycle', *Photosynthesis Research* 140 (2019) 235–252. An excellent, detailed scientific outline of the whole period of discovery with a balanced final analysis of the rift between Calvin and Benson. While noting that Benson was not blameless in the rift, Sharkey concurs that 'He didn't have to do that. He could have done it right.'

J. A. Bassham, 'Mapping the carbon reduction cycle: a personal retrospective', *Photosynthesis Research* 76 (2003) 35–52. James Bassham was instrumental in completing what might rightly be called the Calvin–Benson–Bassham cycle, were it not such a mouthful. He, too, told the story in *Photosynthesis Research*.

Photophosphorylation

D. I. Arnon, F. R. Whatley and M. B. Allen, 'Photosynthesis by isolated chloroplasts. II. Photosynthetic phosphorylation, the conversion of light into phosphate bond energy', *Journal of the American Chemical Society* 76 (1954) 6324–6329. The paper that made Daniel Arnon's name; the title says it all. Published in the same volume as the paper that led to Calvin's Nobel prize.

Reverse Krebs cycle

K. V. Thimann, 'The absorption of carbon dioxide in photosynthesis', *Science* 88 (1938) 506–507. A beautiful short hypothesis from Kenneth Thimann, who more or less predicts the reverse Krebs cycle as the mechanism of CO_2 fixation in photosynthesis, just a year after Krebs proposed his cycle.

Further reading

D. I. Arnon, 'Ferredoxin and photosynthesis', *Science* 149 (1965) 1460–1470. Ferredoxin was discovered in the early 1960s and in this paper Arnon brought together an impressive set of experimental observations, ending with the teaser 'Evidence for the participation of bacterial ferredoxin in bacterial photophosphorylation is being actively sought.' The reverse Krebs cycle followed the next year.

M. C. Evans, B. B. Buchanan and D. I. Arnon, 'A new ferredoxin-dependent carbon reduction cycle in a photosynthetic bacterium', *Proceedings of the National Academy of Sciences USA* 55 (1966) 928–934. An all-time classic paper, which somehow remained controversial or ignored over the following quarter of a century. Just beautiful science.

B. B. Buchanan, R. Sirevåg, G. Fuchs, R. N. Ivanovsky, Y. Igarashi, M. Ishii, F. R. Tabita and I. A. Berg, 'The Arnon–Buchanan cycle: a retrospective, 1966–2016', *Photosynthesis Research* 134 (2017) 117–131. A fascinating compendium of recollections from the main protagonists of the reverse Krebs cycle, fifty years after it was first proposed – by which time it is at last undisputed, if not widely known.

B. B. Buchanan, 'Daniel I. Arnon. November 14, 1910–December 20, 1994', *Biographical Memoirs* 80 (2001) 2–20. A brief biography of Daniel Arnon (in the US National Academy of Sciences collection) from his friend and long-term colleague Bob Buchanan.

B. B. Buchanan, 'Thioredoxin: an unexpected meeting place', *Photosynthesis Research* 92 (2007) 145–148. Bob Buchanan's account of the difficult relationship between Calvin and Arnon, and how Andrew Benson later bridged the intellectual gap.

The problems with peer review

Don Braben, *Scientific Freedom: The Elixir of Civilization* (San Francisco, Stripe Press, 2020). A fine book that makes the case for rethinking how science is funded. Peer review is deeply conservative and tends to oppose toppling the status quo. Radical new ideas do exactly this. The question is how best to identify and support radically new ideas. Braben fleshes out the problem through a series of potted biographies in another book whose subtitle hints at the ambition: *Promoting the Planck Club: How Defiant Youth, Irreverent Researchers and Liberated Universities Can Foster Prosperity Indefinitely* (Hoboken, NJ, John Wiley, 2014).

Ferredoxin and photorespiration

J. Ormerod, '"Every dogma has its day": a personal look at carbon metabolism in photosynthetic bacteria', *Photosynthesis Research* 76 (2003) 135–143. Yes, I stole his phrase. A very personable account from an experimentalist and deep thinker, who started out doing manometry at Sheffield University, in the same department where Krebs had worked, before moving to Norway for the rest of his career. Here, Ormerod proposes that photorespiration is useful because it oxidises ferredoxin, lowering oxidative stress.

Y. Shomura, M. Taketa, H. Nakashima *et al.*, 'Structural basis of the redox switches in the NAD^+-reducing soluble [NiFe]-hydrogenase', *Science* 357 (2017) 928–932. A neat paper, which shows that an H_2-oxidising bacterium can survive exposure to oxygen by switching its hydrogenase enzymes to an inactive conformation, protecting them from reactive oxygen species. It's possible … but this paper just highlights how difficult it is to operate the reverse Krebs cycle when oxygen is present.

Widespread reverse Krebs cycle

A. Mall, J. Sobotta, C. Huber *et al.*, 'Reversibility of citrate synthase allows autotrophic growth of a thermophilic bacterium', *Science* 359 (2018) 563–567. A seminal discovery, that the canonical ATP citrate lyase is not necessary for the reverse Krebs cycle; the standard enzyme citrate synthase is reversible after all. So do dogmas fall.

L. Steffens, E. Pettinato, T. M. Steiner, A. Mall, S. König, W. Eisenreich and I. A. Berg, 'High CO_2 levels drive the TCA cycle backwards towards autotrophy', *Nature* 592 (2021) 784–788. Another paper from Ivan Berg's group, this time showing that high atmospheric CO_2 (as on the early Earth) tends to drive the normal oxidative cycle in reverse, to fix CO_2. From being restricted to a tight phylogenetic corner, the reverse Krebs cycle now looks like it could be widespread in bacteria and archaea.

I. A. Berg, 'Ecological aspects on the distribution of different autotrophic CO_2 fixation pathways', *Applied and Environmental Microbiology* 77 (2011) 1925–1936. One of the pioneers of alternative CO_2 fixation pathways reflects on their distribution and ecology; now a little dated by his own recent discoveries, but still full of insight.

Chapter 3: From gases to life

Discovery of black smoker vents

J. B. Corliss, J. A. Baross and S. . Hoffman, 'An hypothesis concerning the relationship between submarine hot springs and the origin of life on earth', *Oceanologica Acta* Special Issue (1981) 0399-1784. Hot on the heels of the discovery of black smoker vents in 1979 came this first conceptualisation of how vents might have the primordial setting for life's origin.

J. A. Baross and S. E. Hoffman, 'Submarine hydrothermal vents and associated gradient environments as sites for the origin and evolution of life', *Origins of Life and Evolution of Biospheres* 15 (1985) 327–345. A serious wrestling with how steep chemical gradients in hydrothermal environments could promote the origins of biochemistry.

Discovering Hydrothermal Vents. A wealth of interesting information on the discovery of black smokers on the Woods Hole Oceanographic Institute: https://www.whoi.edu/feature/history-hydrothermal-vents/index.html. WHOI commissioned the *Alvin* submersible in 1964, and have maintained it over nearly 60 years. I usually try to avoid giving web addresses, in case they disappear, but this material is hard to beat.

Reanalysis of early Earth hydrothermal systems

W. Martin, J. Baross, D. Kelley and M. J. Russell, 'Hydrothermal vents and the origin of life', *Nature Reviews Microbiology* 6 (2008) 805–814. A good overview of the different types of vent system and their roles at the origin of life, from two distinguished marine biologists, John Baross and Deb Kelley, who had each discovered new vent systems, alongside Bill Martin and Mike Russell, leading protagonists of how life might have started in such systems.

N. H. Sleep, D. K. Bird and E. C. Pope, 'Serpentinite and the dawn of life', *Philosophical Transactions of the Royal Society B* 366 (2011) 2857–2869. An excellent article that considers how the geological record constrains hydrothermal theories for the origin of life.

N. T. Arndt and E. G. Nisbet, 'Processes on the young Earth and the habitats of early life', *Annual Review of Earth and Planetary Sciences* 40 (2012) 521–49. Gives a good sense of how geologists' understanding of the Hadean has shifted in recent decades; no longer is the early Earth thought of as a hellhole of boiling magma and asteroid impacts, but rather as a relatively tranquil waterworld.

F. Westall, K. Hickman-Lewis, N. Hinman, P. Gautret, K. A. Campbell, J. G. Breheret, F. Foucher, A. Hubert, S. Sorieul, A. V. Dass, T. P. Kee, T. Georgelin and A. Brack, 'A hydrothermal-sedimentary context for the origin of life', *Astrobiology* 18 (2018) 259–293. A well-balanced overview of geological conditions on the early Earth in relation to life's origins.

Mike Russell and Gunter Wächtershäuser

G. Wächtershäuser, 'Evolution of the first metabolic cycles', *Proceedings of the National Academy of Sciences USA* 87 (1990) 200–204. One of the first papers to take the reverse Krebs cycle seriously as prebiotic chemistry, giving a detailed exposition of how it might occur on mineral surfaces, pulled through by the formation of iron sulfides, notable iron pyrites (fool's gold). Unusually philosophical in its approach, enumerating a number of postulates and theorems, deriving from Popper's ideas on testing hypotheses. A seminal paper.

G. Wächtershäuser, 'Groundworks for an evolutionary biochemistry: the iron–sulphur world', *Progress in Biophysics and Molecular Biology* 58 (1992) 85–201. An exhaustive, revolutionary thesis, detailing much of Wächtershäuser's early thinking on an autotrophic origin of life. More of a whole book than a paper; justly famous.

M. J. Russell and A. J. Hall, 'The emergence of life from iron monosulphide bubbles at a submarine hydrothermal redox and pH front', *Journal of the Geological Society* 154 (1997) 377–402. Mike Russell and Allan Hall had been writing papers on alkaline hydrothermal vents for nearly a decade at this point, but to my mind this is the most complete early formulation, which incorporated their detailed thinking on Peter Mitchell's protonmotive force and the reverse Krebs cycle.

W. Martin and M. J. Russell, 'On the origins of cells: a hypothesis for the evolutionary transitions from abiotic geochemistry to chemoautotrophic prokaryotes, and from prokaryotes to nucleated cells', *Philosophical Transactions of the Royal Society B* 358 (2003) 59–85. An all-time classic paper, the kind of paper that draws young minds not just into the field but into science.

J. Whitfield, 'Origin of life: nascence man', *Nature* 459 (2009) 316–319. Rarely does *Nature* feature the work of a scientist so explicitly, in this case even including a photoshopped image of Russell as the renaissance man Erasmus. Quite a tribute.

Discovery of Lost City

D. S. Kelley, J. A. Karson, D. K. Blackman, G. L. Früh-Green, D. A. Butterfield, M. D. Lilley, E. J. Olson, M. O. Schrenk, K. K. Roe, G. T. Lebon, P. Rivizzigno; AT3-60 Shipboard Party, 'An off-axis hydrothermal vent field near the Mid-Atlantic Ridge at 30 degrees N', *Nature* 412 (2001) 145–149. An extraordinary discovery, which has inspired whole fields of research and demonstrates the value of exploration and discovery science.

D. S. Kelley, J. A. Karson, G. L. Früh-Green *et al.*, 'A serpentinite-hosted ecosystem: the Lost City hydrothermal field,' *Science* 307

(2005) 1428–1434. More on Lost City from Deborah Kelley and colleagues, linking the geology to the chemistry and microbiology of this iconic system.

D. S. Kelley, 'From the mantle to microbes: the Lost City hydrothermal field', *Oceanography* 18 (2005) 32–45. A personal piece by Deborah Kelley, who captained the *Alvin* submersible on that serendipitous discovery of Lost City in 2000. Richly illustrated and with an exciting early view of the significance of serpentinisation for life elsewhere in the solar system.

Terrestrial geothermal systems

David W. Deamer, *Assembling Life: How Can Life Begin on Earth and Other Habitable Planets?* (Oxford, Oxford University Press, 2019). David Deamer has worked on the origin of life over several decades and did pioneering work on lipid membranes and protocells, including work on a 'lipid world' with Harold Morowitz. He was one of the developers of the nanopore technology for DNA sequencing, which relies on membrane potential. In recent years, he has become a strong but open-minded proponent of terrestrial geothermal systems. I don't agree with him about this, but we have had many constructive discussions. This book reflects his desire to explicitly test the predictions of competing hypotheses, with which I wholeheartedly concur.

A. Y. Mulkidjanian, A. Y. Bychkov, D. V. Dibrova, M. Y. Galperin, E. V. Koonin, 'Origin of first cells at terrestrial, anoxic geothermal fields', *Proceedings of the National Academy of Sciences USA* 109 (2012) E821–830. Armen Mulkidjanian is a distinguished bioenergeticist who has had some strikingly original ideas on the origin of life involving an unusual form of 'zinc sulfide photosynthesis' that would have had to take place in high-pressure atmospheres. Here he teams up with some leading phylogeneticists including the brilliant and prolific Eugene Koonin. I am unpersuaded, but Mulkidjanian's papers are always worth reading.

J. D. Sutherland, 'Studies on the origin of life: the end of the beginning', *Nature Reviews in Chemistry* 1 (2017) 0012. John Sutherland is a prominent chemist who has developed an impressive network of prebiotic chemistry to synthesise many of the core building blocks of life, including nucleotides, in what he calls 'cyano-sulfidic protometabolism'. The downside is that this chemistry does not look much like biochemistry. Does that matter? In my view it does, but until we know more, that is a matter of subjective judgement. Sutherland expressed his own judgement in this article: 'the idea that life originated at vents should, like the vents themselves, remain "In the deep bosom of the ocean buried."' If I agreed with him, I would not have written this book.

J. Szostak, 'How did life begin?', *Nature* 557 (2018) S13 –S15. Jack Szostak won the Nobel prize for his work on telomeres and since then has focused on the origin of life, producing many fine papers. He works closely with John Sutherland on prebiotic chemistry and has little sympathy with the idea of vents. This article includes the line: 'Strikingly, many of the chemical intermediates on the way to RNA crystallize out of reaction mixtures, self-purifying and potentially accumulating on the early Earth as organic minerals – reservoirs of material waiting to come to life when conditions change.' This conception is plainly far from the continuous growth that I advocate in this book. I must ask: what does that last line really mean?

Protometabolism using life as a guide

Christian De Duve, *Singularities. Landmarks on the Pathways of Life* (Cambridge, Cambridge University Press, 2005). I didn't mention de Duve's thinking much in this book but I should have done. Like Morowitz, de Duve focused on how geochemistry could give rise to biochemistry. His views on the importance of thioesters and energy currencies such as acetyl phosphate are beginning to gain experimental support.

M. Preiner, K. Igarashi, K. B. Muchowska, M. Yu, S. J. Varma, K. Kleinermanns, M. K. Nobu, Y. Kamagata, H. Tüysüz, J. Moran and W. F. Martin, 'A hydrogen-dependent geochemical analogue of primordial carbon and energy metabolism', *Nature Ecology and*

Evolution 4 (2020) 534–542. Three different groups came together in this paper to consider the role of iron-containing minerals (greigite, magnetite and awaruite) in catalysing the reaction between H_2 and CO_2 to form reverse-Krebs-cycle intermediates, including in this case acetate and pyruvate. The immediate electron donor here is hydrogen, not raw iron – as happens in life.

S. J. Varma, K. B. Muchowska, P. Chatelain and J. Moran, 'Native iron reduces CO_2 to intermediates and end-products of the acetyl-CoA pathway', *Nature Ecology and Evolution* 2 (2018) 1019–1024. Joseph Moran's group is unusual in that they trained as synthetic chemists, yet are thinking from a biochemical point of view, looking for similar chemistry to that still happening in cells. This is an exciting paper which links up with their own earlier work, showing that various intermediates in the acetyl CoA pathway as well as the reverse Krebs cycle are produced from CO_2 using raw iron as a source of electrons.

K. B. Muchowska, S. J. Varma and J. Moran, 'Synthesis and breakdown of universal metabolic precursors promoted by iron', *Nature* 569 (2019) 104–107. Some wild loops around the Krebs cycle driven by glyoxylate, a two-carbon intermediate that 'short-circuits' the Krebs cycle in plants and some bacteria (the glyoxylate cycle was also discovered by Krebs, working with Hans Kornberg and published in *Nature* in 1957). I'm not sure how close to life's chemistry this really is, but it certainly ropes in a lot of intermediates that are still central to metabolism.

M. Ralser, 'An appeal to magic? The discovery of a non-enzymatic metabolism and its role in the origins of life', *Biochemical Journal* 475 (2018) 2577–2592. An expert on metabolic biochemistry in health, Markus Ralser has published a number of important papers on how metabolic pathways arose before genes, with experimental work showing that glycolysis, the pentose phosphate pathway, gluconeogenesis and parts of the Krebs cycle can all operate as spontaneous chemistry, albeit generally (in his experiments) in the direction of degeneration rather than synthesis, and at very low concentrations of intermediates. This paper summarises Ralser's work, including his thinking on why metabolic pathways had to evolve before genes. On that I think he's just right.

Michael Madigan, Kelly Bender, Daniel Buckley, W. Matthew Sattley and David Stahl, *Brock Biology of Microorganisms*, 15th edn (London, Pearson, 2018). One of the few microbiology textbooks that grapples in a serious way with the origin of life. Given the parallels between vent chemistry and microbial biochemistry, it is not surprising that they favour hydrothermal origins.

H. Hartman, 'Speculations on the origin and evolution of metabolism', *Journal of Molecular Evolution* 4 (1975) 359–70. An old but still penetrating paper, one of the first to think through metabolic chemistry starting from CO_2 in terms of the Krebs cycle, not so long after the early papers on the reverse Krebs cycle.

The physiology of LUCA

M. C. Weiss, F. L. Sousa, N. Mrnjavac, S. Neukirchen, M. Roettger, S. Nelson-Sathi and W. F. Martin, 'The physiology and habitat of the last universal common ancestor', *Nature Microbiology* 1 (2016) 16116. A controversial paper that attempts to tack around the knotty problem of lateral gene transfer to advance a striking portrait of the physiology of LUCA – as a population of simple cells subsisting in vents in the neverland between the abiotic and biotic. While some details have been decried, the paper reinforces the idea that LUCA lived from H_2 and CO_2 using proton gradients and iron–sulfur proteins.

R. Braakman and E. Smith, 'The emergence and early evolution of biological carbon fixation', *PLoS Computational Biology* 8 (2012) e1002455. An extremely neat idea, which attempts to unite phylogenetic analysis with metabolic flux analysis to constrain the deepest branches of the tree of life using the 'rule' that cells must always be able to grow, and therefore must have had a functional metabolic network. Simple in theory, but difficult in practice, and some of their conclusions stretch credulity. But surely the future lies in the type of analysis pioneered in this study.

F. L. Sousa, T. Thiergart, G. Landan, S. Nelson-Sathi, I. A. C. Pereira, J. F. Allen, N. Lane and W. F. Martin, 'Early bioenergetic evolution', *Philosophical Transactions of the Royal Society B* 368 (2013) 20130088. A detailed paper that attempts to constrain the earliest steps of evolution in terms of the machinery of energy transduction.

Acetyl CoA pathway

M. J. Russell and W. Martin, 'The rocky roots of the acetyl-CoA pathway', *Trends in Biochemical Sciences* 29 (2004) 358–363. A little gem of a paper, which points to the ubiquity of metal ion clusters, notably iron–sulfur clusters, in the most ancient metabolic pathways.

G. Fuchs, 'Alternative pathways of carbon dioxide fixation: insights into the early evolution of life?', *Annual Review of Microbiology* 65 (2011) 631–658. There are only six known pathways of CO_2 fixation, and this authoritative and detailed paper from Georg Fuchs (who had a hand in discovering three of them) considers what they say about the earliest stages of life. In essence, they point to the centrality of both acetyl CoA and ferredoxin.

W. Nitschke, S. E. McGlynn, J. Milner-White and M. J. Russell, 'On the antiquity of metalloenzymes and their substrates in bioenergetics', *Biochimica et Biophysica Acta Bioenergetics* 1827 (2013) 871–881. Wolfgang Nitschke trained as a physicist before turning to molecular biology. He has profound insights into the workings of redox proteins and has teamed up with Mike Russell on their emergence at the origin of life.

Harold Morowitz and the reverse Krebs cycle

Harold J. Morowitz, *Energy Flow in Biology* (New York, Academic Press, 1968). A classic work, where Morowitz introduces the idea that energy flows, matter cycles. Difficult and mathematical but worthwhile.

Eric Smith and Harold J. Morowitz, *The Origin and Nature of Life on Earth: The Emergence of the Fourth Geosphere* (Cambridge, Cambridge University Press, 2016). A magisterial volume, and fitting final testament of Morowitz, written with his long-term collaborator, the equally brilliant Eric Smith. Deserves a place on the shelf of anyone seriously interested in the origin of life. Manages to present new ideas with admirable clarity, while being balanced and exhaustive.

J. Trefil, H. J. Morowitz and E. Smith, 'The origin of life. A case is made for the descent of electrons', *American Scientist* 206 (2009) 96–213. A fine short piece that is serious but accessible, with an enjoyable title that puns on thermodynamics and Darwin's *Descent of Man*.

Leslie Orgel and an appeal to magic

L. Orgel, 'Self-organizing biochemical cycles', *Proceedings of the National Academy of Sciences USA* 97 (2000) 12503–12507. Orgel wrote this paper as a direct refutation of Morowitz, describing the idea of self-organising cycles as 'an appeal to magic'. He warmed to this theme in his final (posthumous) paper, cited in my Introduction, describing such cycles as 'if pigs could fly chemistry' in *PLoS Biology* 6 (2008) e18.

Nucleotide synthesis using life as a guide

S. A. Harrison and N. Lane, 'Life as a guide to prebiotic nucleotide synthesis', *Nature Communications* 9 (2018) 5176–5177. Stuart Harrison did his PhD in my lab and is now working with me as a post-doc. This little paper laid out our ideas for how nucleotides might be synthesised by following the biological pathway, using metal ions as catalysts. Since then he successfully synthesised the nucleobase uracil this way, but we are only writing up this work now so I can't cite it here.

Magic surfaces

G. D. Cody, N. Z. Boctor, R. M. Hazen, J. A. Brandes, H. J. Morowitz, H. S. Yodor Jr, 'Geochemical roots of autotrophic carbon fixation: hydrothermal experiments in the system citric acid, H_2O-($\pm$FeS)-($\pm$NiS)', *Geochimica et Cosmochimica Acta* 65 (2001) 3557–3576.

Early work on the propensity of iron-sulfide minerals to catalyse the chemistry of the Krebs cycle from George Cody and colleagues including Morowitz.

E. Camprubi, S. F. Jordan, R. Vasiliadou and N. Lane, 'Iron catalysis at the origin of life', *IUBMB Life* 69 (2017) 373–381. A theoretical chemistry paper that details the reverse Krebs cycle (or rather, line) as we imagine it occurring across the surface of semiconducting iron–nickel–sulfur minerals.

Methanogens as a guide

R. K. Thauer, A.-K. Kaster, H. Seedorf, W. Buckel and R. Hedderich, 'Methanogenic archaea: ecologically relevant differences in energy conservation', *Nature Reviews Microbiology* 6 (2008) 579–591. A treasure-trove of data and mechanisms on the energetics of making a living at the margins of the thermodynamically permissible.

V. Sojo, B. Herschy, A. Whicher, E. Camprubi and N. Lane, 'The origin of life in alkaline hydrothermal vents', *Astrobiology* 16 (2016) 181–197. Our own attempt to reconcile the origins of life in alkaline hydrothermal vents with the divergence of bacteria and archaea, based on electron bifurcation and the origins of pumping. We conceived LUCA as being essentially similar to methanogens, relying on proton gradients to fix CO_2 using the energy-converting hydrogenase and ferredoxin.

pH gradients

N. Lane, J. F. Allen and W. Martin, 'How did LUCA make a living? Chemiosmosis in the origin of life', *BioEssays* 32 (2010) 271–280. Wrestling with the problems of how and when chemiosmotic coupling might have arisen in early cells. In retrospect I think we focused too much attention on ATP, and not enough on CO_2 fixation, but that's a bone of contention.

N. Lane, 'Why are cells powered by proton gradients?', *Nature Education* 3 (2010) 18. A concise account of the problem and possible solutions, as part of a series of really valuable short articles aimed at students and a general scientific readership. Do take a look at *Scitable*, home of this educational resource.

N. Lane and W. Martin, 'The origin of membrane bioenergetics', *Cell* 151 (2012) 1406–1416. A paper with some important ideas, if I say so myself, forged in the exciting crucible of discussion with Bill Martin. I'm proud of this paper.

B. Herschy, A. Whicher, E. Camprubi, C. Watson, L. Dartnell, J. Ward, J. R. G. Evans and N. Lane, 'An origin-of-life reactor to simulate alkaline hydrothermal vents', *Journal of Molecular Evolution* 79 (2014) 213–227. Our early attempts to build an 'origin-of-life' reactor, many of which failed. This was a mixed bag of a paper that included some of the highlights of a system that ultimately had too many variables. Now we're using microfluidics.

N. Lane, 'Proton gradients at the origin of life', *BioEssays* 39 (2017) 1600217. A riposte to the criticisms of Baz Jackson, a fine bioenergeticist, who sadly died of cancer soon afterwards. Towards the end of his life, he took an interest in the origins of life in alkaline hydrothermal vents, raising a number of objections. I disagreed, but felt we were not quite understanding each other's terms of reference. I regret never meeting him in person to discuss these questions.

R. Vasiliadou, N. Dimov, N. Szita, S. Jordan and N. Lane, 'Possible mechanisms of CO_2 reduction by H_2 via prebiotic vectorial electrochemistry', *Royal Society Interface Focus* 9 (2019) 20190073. A detailed theoretical paper with some experimental results, showing that protons can cross iron–sulfur barriers two million times faster than hydroxide ions going the other way, giving rise to very steep pH gradients (as much as 3 or 4 pH units across single nanocrystals).

R. Hudson, R. de Graaf, M. S. Rodin, A. Ohno, N. Lane, S. E. McGlynn, Y. M. A. Yamada, R. Nakamura, L. M. Barge, D. Braun and V. Sojo, 'CO_2 reduction driven by a pH gradient', *Proceedings of the National Academy of Sciences USA* 117 (2020) 22873–22879. The first robust demonstration that proton gradients across semiconducting barriers really can promote the transfer of electrons from

H_2, on one side, to CO_2 on the other, to form organic molecules (in this case, formate).

Formation of protocells in hydrothermal vents

T. M. McCollom, G. Ritter and B. R. Simoneit, 'Lipid synthesis under hydrothermal conditions by Fischer–Tropsch-type reactions', *Origins of Life and Evolution of Biospheres* 29 (1999) 153–166. Starting with formate at high pressure and high temperature, McCollom and colleagues succeeded in producing long-chain hydrocarbons, fatty acids and alcohols relevant to the origin of life. Curiously, it only worked in steel (not glass) reactors.

S. F. Jordan, H. Rammu, I. Zheludev, A. M. Hartley, A. Marechal and N. Lane, 'Promotion of protocell self-assembly from mixed amphiphiles at the origin of life', *Nature Ecology & Evolution* 3 (2019) 1705–1714. Rather surprisingly, we found that alkaline hydrothermal conditions actually favour the spontaneous assembly of 'protocells' (with lipid bilayer membranes enclosing an aqueous space) from simple mixtures of prebiotically plausible fatty acids and fatty alcohols.

S. F. Jordan, E. Nee and N. Lane, 'Isoprenoids enhance the stability of fatty acid membranes at the emergence of life potentially leading to an early lipid divide', *Royal Society Interface Focus* 9 (2019) 20190067. Amazingly, protocells with membranes composed mostly of fatty acids (like bacteria) tend to stick to mineral surfaces, whereas those incorporating isoprenoids (like archaea) do not. Something to do with curvature.

S. F. Jordan, I. Ioannou, H. Rammu, A. Halpern, L. K. Bogart, M. Ahn, R. Vasiliadou, J. Christodoulou, A. Maréchal and N. Lane, 'Spontaneous assembly of redox-active iron–sulfur clusters at low concentrations of cysteine', *Nature Communications* 12 (2021) 5925. Our own demonstration that the biological iron–sulfur clusters found in ferredoxin, which promote CO_2 fixation in cells, can form spontaneously under prebiotic conditions.

T. West, V. Sojo, A. Pomiankowski and N. Lane, 'The origin of heredity in protocells', *Philosphical Transactions of the Royal Society B* 372 (2017) 20160419. I think this is an important paper, but it has stayed under the radar. We show that positive feedbacks between organic molecules (fatty acids and amino acids) and iron–sulfur clusters drive a form of membrane heredity that makes it possible for protocells to get better at copying themselves. The basis for much of our own current thinking on the origins of the genetic code.

R. Nunes Palmeira, M. Colnaghi, S. Harrison, A. Pomiankowski and N. Lane, 'The limits of metabolic heredity in protocells', available on *BioRxiv* (https://doi.org/10.1101/2022.01.28.477904). More on positive feedbacks and autocatalysis in protocells. Only general forms of catalysis are helpful, as they keep a balance between the different protometabolic pathways needed for protocell growth.

Energy flow in hydrothermal vents

J. P. Amend and T. M. McCollom, 'Energetics of biomolecule synthesis on early Earth', in L. Zaikowski *et al.* (eds), *Chemical Evolution II: From the Origins of Life to Modern Society* (American Chemical Society, 2009) pp. 63–94. I still find this mind-blowing. Which is more stable, thermodynamically, a mixture of H_2 and CO_2, or cells? The answer is cells. That's why life exists.

J. P. Amend, D. E. LaRowe, T. M. McCollom and E. L. Shock, 'The energetics of organic synthesis inside and outside the cell', *Philosophical Transactions of the Royal Society B* 368 (2013) 20120255. More on the thermodynamics of prebiotic chemistry in hydrothermal systems from Amend and McCollom, with 'the godfather', Everett Shock, who came up with the memorable quip that living on H_2 and CO_2 is a 'free lunch that you're paid to eat'.

W. F. Martin, F. L. Sousa and N. Lane, 'Energy at life's origin', *Science* 344 (2014) 1092–1093. A short take on why life itself points to hydrothermal vents as the site of its origin. Occam's razor.

A. Whicher, E. Camprubi, S. Pinna, B. Herschy and N. Lane, 'Acetyl phosphate as a primordial energy currency at the origin of life', *Origins of Life and Evolution of Biospheres* 48 (2018) 159–179. Our first experiments on acetyl phosphate as a prebiotic energy currency. It works quite well at phosphorylating sugars, but can play havoc by acetylating amino acids.

S. Pinna, C. Kunz, S. Harrison, S. F. Jordan, J. Ward, F. Werner and N. Lane, 'A prebiotic basis for ATP as the universal energy currency', *BioRxiv* https://doi.org/10.1101/2021.10.06.463298 (2021). Amazingly, acetyl phosphate is unique in that it will phosphorylate ADP to ATP in water, but can't do this with other nucleoside diphosphates. This suggests that ATP became established as the universal energy currency because of its unusual prebiotic chemistry.

Chapter 4: Revolutions

Cambrian explosion

Stephen Jay Gould, *Wonderful Life: The Burgess Shale and the History of Nature* (New York, W. W. Norton, 1989). A heady book that must have drawn a generation of young minds into palaeontology. But it has not weathered so well, and pushes its central thesis a little too far.

Daniel C. Dennett, *Darwin's Dangerous Idea: Evolution and the Meanings of Life* (New York, Simon & Schuster, 1995). I never appreciated what a razor-sharp philosophical mind could do to scientific arguments until I read this book, where Dan Dennett skewers some of the biggest names in biology, including Stephen Jay Gould's take on the Cambrian explosion. Revelatory.

M. A. S. McMenamin, 'Cambrian chordates and vetulicolians', *Geosciences* 9 (2019) 354. A useful update on the early Cambrian chordates, which frustratingly lacks even estimated dates (you'll have to look into the references cited).

J. Y. Chen, D. Y. Huang and C. W. Li, 'An early Cambrian craniate-like chordate', *Nature* 402 (1999) 518–522. A flavour of new discoveries of early chordates in the Maotianshan Shale in China; observes coyly that 'These findings will add to the debate on the evolutionary transition from invertebrate to vertebrate.'

Further reading

S. Conway Morris, 'Darwin's dilemma: the realities of the Cambrian "explosion"', *Philosophical Transactions of the Royal Society B* 361 (2006) 1069–1083. An enjoyably breezy analysis of the Cambrian explosion from one of the leading protagonists, who came to fame through Gould's *Wonderful Life*, and wrote a slightly resentful book to set a few matters straight, entitled *The Crucible of Creation* (Oxford, Oxford University Press, 1998).

Oxygen

Donald E. Canfield, *Oxygen: A Four Billion Year History* (Princeton, Princeton University Press, 2014). A fine book by one of the leading protagonists of early-Earth history, rolling up his sleeves on how we can know anything at all about the composition of the air and oceans billions of years ago.

Nick Lane, *Oxygen: The Molecule that Made the World* (Oxford, Oxford University Press, 2002). My own take on the evolutionary history of oxygen, written in 2002. Given how much has changed in two decades it hasn't weathered badly, though I no longer agree with all that I wrote. (Others do and think I'm wrong now.) The material remains vivid and fascinating.

Andrew H. Knoll, *Life on a Young Planet: The First Three Billion Years of Life on Earth* (Princeton, Princeton University Press, 2003). An authoritative, personable and engaging first-hand take on the difficulty of knowing anything about the world as it was billions of years ago. Andy Knoll combines geology and biology with unusual insight and is one of the most respected and likeable voices in palaeontology.

N. J. Butterfield, 'Oxygen, animals and oceanic ventilation: an alternative view', *Geobiology* 7 (2009) 1–7. Quite a diatribe against the lazy ascription of dramatic environmental change to the simple gas oxygen. A valuable corrective.

O. Judson, 'The energy expansions of evolution', *Nature Ecology and Evolution* 1 (2017) 0138. A sweeping take from Olivia Judson on how a succession of energy revolutions transformed the potential of life on Earth, from the origins of life to photosynthesis, oxygen and fire.

Electron bifurcation

W. Buckel and R. K. Thauer, 'Flavin-based electron bifurcation, ferredoxin, flavodoxin, and anaerobic respiration with protons (Ech) or NAD$^+$ (Rnf) as electron acceptors: a historical review', *Frontiers in Microbiology* 9 (2018) 401. A clear historical review of one of the most important advances in bioenergetics, nay, in all microbiology, by two great pioneers of the field who had been wrestling with the problem for the best part of half a century.

Margaret Dayhoff and Lynn Margulis

M. O. Dayhoff, R. V. Eck, E. R. Lippincott and C. Sagan, 'Venus: atmospheric evolution', *Science* 155 (1967) 556–558. One of Margaret Dayhoff's collaborations with Carl Sagan, calculating the equilibrium composition of Venus's atmosphere, and ruling out the possibility of even small organic molecules surviving there. Still relevant today in light of the disputed detection of phosphine gas in the Venusian atmosphere.

R. V. Eck and M. O. Dayhoff, 'Evolution of the structure of ferredoxin based on living relics of primitive amino acid sequences', *Science* 152 (1966) 363–366. One of my favourite papers, an utter masterpiece of scientific deduction.

R. M. Schwartz and M. O. Dayhoff, 'Origins of prokaryotes, eukaryotes, mitochondria, and chloroplasts. A perspective is derived from protein and nucleic acid sequence data', *Science* 199 (1978) 395–403. Margaret Dayhoff getting into her stride, dealing with the origin of just about everything that matters in the world of prokaryotes.

J. Barnabas, R. M. Schwartz and M. O. Dayhoff, 'Evolution of major metabolic innovations in the Precambrian', *Origins of Life* 12 (1982) 81–91. A late classic from Margaret Dayhoff, sketching the deep roots of the tree of life, and supporting the inferences of Lynn Margulis on bacterial ancestry of mitochondria and chloroplasts.

L. Sagan, 'On the origin of mitosing cells', *Journal of Theoretical Biology* 14 (1967) 225–274. A wonderful paper that reconceptualised the evolution of cells on a planetary scale. Revelatory and justly famous, if not correct about everything (as is normal in science). Note that Lynn Margulis had recently divorced Carl Sagan, hence the name Sagan on this paper.

N. Lane, 'Serial endosymbiosis or singular event at the origin of eukaryotes?', *Journal of Theoretical Biology* 434 (2017) 58–67. My own appraisal of Lynn Margulis's contribution to cell evolution, written for a special fiftieth anniversary issue of the journal in which she published her classic 1967 paper on the origin of eukaryotes. An honour to contribute.

Bacterial metabolism and syntrophy

P. Schönheit, W. Buckel and W. F. Martin, 'On the origin of heterotrophy', *Trends in Microbiology* 24 (2016) 12–25. An important paper, which is so clearly thought through that it is almost impossible to see how else heterotrophy ('eating' organic molecules) could have arisen. A little masterpiece.

S. E. McGlynn, G. L. Chadwick, C. P. Kempes and V. J. Orphan, 'Single cell activity reveals direct electron transfer in methanotrophic consortia', *Nature* 526 (2015) 531–535. A beautiful paper harnessing stunning technology, which shows that tight-knot consortia of cells form clumps with regular numbers of specific types of cell (an optimal stoichiometry), which transfer electrons between cells.

E. Libby, L. Hébert-Dufresne, S.-R. Hosseini and A. Wagner, 'Syntrophy emerges spontaneously in complex metabolic systems', *PLoS Computational Biology* 15 (2019) e1007169. A clever paper showing how metabolic syntrophy (mutual reliance) can arise simply through mutations causing the partial degeneration of individual cells.

Paul G. Falkowski, *Life's Engines: How Microbes made the Earth Habitable* (Princeton, Princeton University Press, 2015). An engaging take on the history of life from the point of view of life's engines – both the protein machines and the bacterial consortia that power evolution. Aimed at the general public by a leading protagonist, there's plenty of good material here on photosynthesis and symbioses between diverse groups.

Dawn of photosynthesis

J. Allen and W. Martin, 'Out of thin air', *Nature* 445 (2007) 610–612. A sparkling piece that elaborates on John Allen's 'redox switch' hypothesis for how the Z-scheme that enabled oxygenic photosynthesis came to be.

Nick Lane, *Life Ascending: The Ten Great Inventions of Evolution* (London, Profile Books and New York, W. W. Norton, 2009). If you're confused by my references to the Z-scheme here, then try my chapter on photosynthesis in *Life Ascending*, where I devote a chapter to its evolution, focusing on John Allen's 'redox switch' hypothesis.

Nick Lane, *Building with Light: Primo Levi, Science and Writing* (Centro internazionale di studi Primo Levi, 2012). Available here: www.primolevi.it/en/primo-levi-science-writer. An unusual story, told in the piece itself, at the invitation of Levi's son, Renzo. Primo Levi was a wonderful writer and human being, and his book *The Periodic Table* should be read by everyone.

Tim Lenton and Andrew Watson, *Revolutions that Made the Earth* (Oxford, Oxford University Press, 2013). A detailed overview of how the Earth system (or Gaia) can flip from one stable state to another as conditions change, sometimes abruptly. The intellectual son and grandson of James Lovelock (using PhD mentorship as academic lineage), Lenton and Watson know exactly whereof they speak.

H. C. Betts, M. N. Puttick, J. W. Clark, T. A. Williams, P. C. J. Donoghue and D. Pisani, 'Integrated genomic and fossil evidence illuminates life's early evolution and eukaryote origin', *Nature Ecology and Evolution* 2 (2018) 1556–1562. The Bristol group comprises some of the best phylogeneticists in the world today, and this paper uses sophisticated methods to uncover a wealth of detail on early evolution, giving a series of bold conclusions about the timing of critical events such as the evolution of cyanobacteria, eukaryotes, algae, the last universal common ancestor (LUCA) and the crown groups of bacteria and archaea.

T. Oliver, P. Sánchez-Baracaldo, A. W. Larkum, A. W. Rutherford and T. Cardona, 'Time-resolved comparative molecular evolution of oxygenic photosynthesis', *Biochimica et Biophysica Acta Bioenergetics* 1862 (2021) 148400. I can't help but cite this radically different take on the early origin of oxygenic photosynthesis by the fearless Tanai Cardona and some distinguished colleagues including Patricia Sánchez-Baracaldo. I think it's unlikely they are right about this, but if they are then the foundations of early evolution will shudder. That is always exciting, and science needs more challenging thinking like this.

Shurum conundrum

G. A. Shields, 'Carbon and carbon isotope mass balance in the Neoproterozoic Earth system', *Emerging Topics in Life Sciences* 2 (2018) 257–265. Mass balance might sound dull but if we can't add up then nothing makes sense. Graham Shields uses it as the basis for an important reanalysis of what factors control the accumulation of oxygen in the atmosphere.

G. A. Shields, B. J. W. Mills, M. Zhu, T. D. Raub, S. J. Daines and T. M. Lenton 'Unique Neoproterozoic carbon isotope excursions sustained by coupled evaporite dissolution and pyrite burial', *Nature Geoscience* 12 (2019) 823–827. Graham Shields's thinking about the role of sulfate in the Neoproterozoic oxygenation and the Cambrian explosion makes some explicit predictions about the burial of iron pyrites, which seem to be true.

S. K. Sahoo, N. J. Planavsky, G. Jiang, B. Kendall, J. D. Owens, X. Wang, X. Shi, A. D. Anbar and T. W. Lyons, 'Oceanic oxygenation events in the anoxic Ediacaran ocean', *Geobiology* 14 (2016) 457–468. A fine-grained geological analysis of the precise timing of oxygenation events during the Neoproterozoic and into the Cambrian. Exquisite detail.

G. Shields-Zhou and L. Och, 'The case for a Neoproterozoic

oxygenation event: geochemical evidence and biological consequences', *GSA Today* 21 (2011) 4–11. A marshalling of the evidence that the rise in oxygen in the late Neoproterozoic was linked to the Cambrian explosion. Not a straightforward case!

Active ventilation in animals

A. H. Knoll, R. K. Bambach, D. E. Canfield and J. P. Grotzinger, 'Comparative earth history and late Permian mass extinction', *Science* 273: 452–457 (1996). A revelatory paper for me, which shows that the survivors of the end-Permian extinction were not random, but predominantly animals that could ventilate their respiratory systems.

S. D. Evans, I. V. Hughes, J. G. Gehling and M. L. Droser, 'Discovery of the oldest bilaterian from the Ediacaran of South Australia', *Proceedings of the National Academy of Sciences USA* 117 (2020) 7845–7850. Trace fossils of small bilaterians (a few millimetres in length) crawling through the mud around the time of the Shurum excursion, between 560 and 550 million years ago.

William F. Martin, Aloysius G. M. Tielens and Marek Mentel, *Mitochondria and Anaerobic Energy Metabolism in Eukaryotes: Biochemistry and Evolution* (Berlin, De Gruyter, 2020). A comprehensive treatment of the biochemistry and physiology of eukaryotes (including animals) that live in the absence or near-absence of oxygen. The base case for any adaptations to rising oxygen.

S. Song, V. Starunov, X. Bailly, C. Ruta, P. Kerner, A. J. M. Cornelissen and G. Balavoine, 'Globins in the marine annelid *Platynereis dumerilii* shed new light on hemoglobin evolution in bilaterians', *BMC Evolutionary Biology* 20 (2020) 165. Shows that the 'Ur-bilaterian' (the common ancestor of all bilaterally symmetrical animals) already had at least five haemoglobin genes, long before the evolution of blood as we know it.

Forked Krebs cycle

Laurence A. Moran, Robert Horton, Gray Scrimgeour and Marc

Perry, *Principles of Biochemistry*, 5th edn (London, Pearson, 2011). One of the few biochemistry textbooks that deals with evolution in a serious way. Larry Moran also writes an excellent, hugely informative and good-humoured blog, *Sandwalk* (named after Darwin's walk at Down House). Strongly recommended; and do take a look at his blog.

C. Da Costa and E. Galembeck, 'The evolution of the Krebs cycle: a promising subject for meaningful learning of biochemistry', *Biochemistry and Molecular Biology Education* 44 (2016) 288–296. A thoughtful and balanced piece that should be read by everyone who ever teaches the Krebs cycle.

D. G. Ryan, C. Frezza and L. A. J. O'Neill, 'TCA cycle signalling and the evolution of eukaryotes', *Current Opinion in Biotechnology* 68 (2021) 72–88. I read this paper after finishing the book and would have said more about it if had I spotted it earlier. Argues that Krebs-cycle intermediates were critical to the cross-talk between the host cells and endosymbionts (proto-mitochondria) that gave rise to eukaryotic cells with large, complex genomes. They have to be right.

L. J. Sweetlove, K. F. Beard, A. Nunes-Nesi, A. R. Fernie and R. G. Ratcliffe, 'Not just a circle: flux modes in the plant TCA cycle', *Trends in Plant Sciences* 15 (2010) 462–470. A lovely analysis of different types of flux through the Krebs cycle. This is one of a series of papers by Lee Sweetlove that considers why plants need fast flux through the Krebs cycle for growth rather than energy. Because plants can generate ATP in their chloroplasts, they often use their mitochondria as biosynthetic organelles.

Chapter 5: To the dark side

Oncogenes and tumour-suppressor genes

Robert A. Weinberg, *One Renegade Cell: The Quest for the Origins of Cancer* (Science Masters Series) (New York, Basic Books, 1998). A classic text, which conveys the selfishness of cancer cells and the multiple genetic hits that turn a normal cell into a renegade cancer cell.

D. Hanahan and R. A. Weinberg, 'The hallmarks of cancer', *Cell* 100 (2000) 57–70 . The kind of paper that gives journals like *Cell* their high 'impact factor': this paper has nearly 40,000 citations, and its successor published eleven years later ('Hallmarks of cancer: the next generation', *Cell* 144 (2011) 646–674) has nearly 60,000. As close to the standard view of cancer as it is possible to get.

Pan-Cancer Analysis of Whole Genomes. 'A collection of research and related content from the ICGC/TCGA consortium on whole-genome sequencing and the integrative analysis of cancer', *Nature* Special (5 February 2020). Mostly open access articles, including a 'flagship article' with the same title; but I can't quite shake the feeling that this is Springer-Nature marketing itself, as it has done with the ENCODE project in the past. This kind of big-data science brooks no contradiction. Is it then still science?

Robin Hesketh, *Betrayed by Nature: The War on Cancer* (New York, Palgrave Macmillan, 2012). An enjoyable book that tackles the science of cancer in a serious but good-humoured way, aimed at those who need to know more. This is as clear an account of the genetic basis of cancer as you are likely to read. Hesketh skips lightly over the metabolic perspective I elaborate in this book, even suggesting that Warburg might be quietly pleased that we now recognise 'metabolic perturbation to be a characteristic of most, if not all, cancers'. I suspect Warburg would have raged against such an insipid statement.

Evidence against mutations driving cancer

P. Rous, 'Surmise and fact on the nature of cancer', *Nature* 183 (1959) 1357–1361. Quite a diatribe against the somatic mutation theory of cancer, written in the same year that Rous received the Nobel prize for his work on viruses as a cause of cancer, and with a similar anger (but a completely different view) to Warburg's papers. Much has changed since then, of course, but his closing line still bites: 'Most serious of all the results of the somatic mutation hypothesis has been its effect on research workers. It acts as a tranquillizer on those who believe in it, and this at a time when every worker should feel goaded now and again by his ignorance of what cancer is.'

S. G. Baker, 'A cancer theory kerfuffle can lead to new lines of research', *Journal of the National Cancer Institute* 107 (2015) dju405. A nicely balanced piece, which begins by citing Peyton Rous before touching on many of lines of evidence not explained by somatic mutations, including nuclear transfer and tissue transplantation. Includes a good quote from the physicist Niels Bohr, who remarked 'How wonderful that we have met with a paradox. Now we have some hope of making progress.'

T. N. Seyfried, 'Cancer as a mitochondrial metabolic disease', *Frontiers in Cell and Developmental Biology* 3 (2015) 43. Another paper that points to the difficulties with the somatic mutation theory of cancer. You don't need to believe the alternative hypothesis to recognise that there are problems with the mainstream view.

A. M. Soto and C. Sonnenschein, 'The somatic mutation theory of cancer: growing problems with the paradigm?', *BioEssays* 26 (2004) 1097–1107. After pointing out a few of the problems, offers an alternative hypothesis dubbed the 'tissue organisation field theory'. None of these ideas are mutually exclusive. The fact that one idea does not explain everything does not mean it is completely wrong.

Robin Holliday, *Understanding Ageing* (Cambridge, Cambridge University Press, 1995). A really fine book that gives an excellent overall view of ageing, including the evidence against an accumulation of somatic mutations driving ageing or age-related diseases such as cancer.

C. A. Rebbeck, A. M. Leroi and A. Burt, 'Mitochondrial capture by a transmissible cancer', *Science* 331 (2011) 303. A bit of an oddity, this paper, and very wacky. Some cancers can be transmitted through bites, especially between dogs, wolves and coyotes; but to take hold, tumours often need to steal mitochondria from their host.

Warburg

H. A. Krebs, 'Otto Heinrich Warburg. 1883–1970', *Biographical Memoirs of Fellows of the Royal Society* 18 (1972) 628–699. I cited this in Chapter 1 as well. Krebs at his most gentlemanly in highlighting his mentor's greatest accomplishments, while finding kind words to allay Warburg's less appealing side.

A. M. Otto, 'Warburg effect(s)—a biographical sketch of Otto Warburg and his impacts on tumor metabolism', *Cancer & Metabolism* 4 (2016) 5. A lively and insightful piece that captures a lot of Warburg's thinking that remains relevant for a modern audience interested in cancer.

Martin D. Kamen, *Radiant Science, Dark Politics. A Memoir of the Nuclear Age* (Berkeley, University of California Press, 1985). I have already cited this in Chapter 2 but it's worth another mention here for Kamen's description of meeting Warburg, on his postwar visit to the United States, at the behest of Robert Emerson, who 'ushered him into the Presence'.

G. M. Weisz, 'Dr Otto Heinrich Warburg – survivor of ethical storms', *Rambam Maimonides Medical Journal* 6 (2015) e0008. An interesting short piece that asks how Warburg survived the Nazi period despite his Jewish ancestry.

John N. Prebble, *Searching for a Mechanism: A History of Cell Bioenergetics* (Oxford, Oxford University Press, 2019). Another fine piece of historical science from John Prebble, including an account of the long dispute between Warburg and David Keilin, which arguably cost the latter a Nobel prize that many felt he deserved. Prebble even finds a concealed slight to Keilin in Warburg's Nobel lecture.

Michael S. Rosenwald, 'Hitler's mother was "the only person he genuinely loved." Cancer killed her decades before he became a monster', *Washington Post*, 20 April 2017. An interesting partial explanation of why Hitler had such a fear of cancer, allowing Warburg to survive in Germany, as well as facilitating the escape of his mother's Jewish doctor, Eduard Bloch.

Govindjee, 'On the requirement of minimum number of four versus eight quanta of light for the evolution of one molecule of oxygen in photosynthesis: a historical note', *Photosynthesis Research* 59 (1999) 249–254. Fascinating history from Govindjee in *Photosynthesis*

Research again, this time of the feud between Emerson and his mentor Warburg. Emerson was right.

E. Höxtermann, 'A comment on Warburg's early understanding of biocatalysis', *Photosynthesis Research* 92 (2007) 121–127. Valuable historical context, giving insight into why Warburg was so hostile to more complex ideas of enzymatic catalysis, both for respiration and photosynthesis. He ended up with delusional tunnel vision. I'm reminded of Frank Harold's words, that scientists should strive for open-minded scepticism.

Warburg and his critics on cancer

Otto Warburg, *Über den Stoffwechsel der Tumoren* (Berlin, Springer, 1926). Translated as *The Metabolism of Tumours* (London, Arnold Constable, 1930). Warburg's first publication on cancer, showing extreme accumulation of lactate in cancer cells, which became the basis for his work on cancer over the following decades.

O. Warburg, 'On the origin of cancer cells', *Science* 123 (1956) 309–314. A searing piece that I quote from in some detail in my chapter. Worth taking a look at the original for the full context.

Otto Heinrich Warburg, *The Prime Cause and Prevention of Cancer* (Würzburg, Konrad Triltsch, 1969). A famous manifesto, following on where his *Science* paper left off. The booklet was developed from a lecture at the Lindau Nobel laureate meeting in 1966, and translated into English with the help of Warburg's long-term collaborator Dean Burk (he of the Lineweaver–Burk plot, for fans of enzyme kinetics). Classic Warburg.

B. Chance, 'Profiles and legacies. Was Warburg right? Or was it that simple?', *Cancer Biology & Therapy* 4 125–(2005) 126. A withering critique of Warburg from Britton Chance, one of the few biochemists who could match Warburg's inventive skill in the laboratory (and an Olympic gold medallist in sailing, to boot). A paper of few, well-aimed words.

S. Weinhouse, 'The Warburg hypothesis fifty years later', *Zeitschrift für Krebsforschung* 87 (1976) 115–126. A detailed critique of Warburg's theory written fifty years later, after Warburg's death. Weinhouse and Chance were arguably more responsible than anyone for the collapse in popularity of Warburg's writings on cancer. Ironically, that led to their rediscovery thirty years later as the metabolic aspects of cancer once again came to the fore.

Reinterpretation of the Warburg effect

M. G. Vander Heiden, L. C. Cantley and C. B. Thompson, 'Understanding the Warburg effect: the metabolic requirements of cell proliferation', *Science* 324 (2009) 1029–1033. A masterful paper, which to my mind captures the key to the Warburg effect, which Warburg himself had missed: growth. For reasons that I discuss in my chapter, mild suppression of respiration with age produces a metabolic rewiring leading to a growth phenotype. We put on weight and are more likely to develop cancer.

P. S. Ward and C. B. Thompson, 'Metabolic reprogramming: a cancer hallmark even Warburg did not anticipate', *Cancer Cell* 21 (2012) 297–308. Craig Thompson again, this time discussing evidence that metabolites such as Krebs-cycle intermediates 'can be oncogenic by altering cell signalling and blocking cellular differentiation', giving altered metabolism 'the status of a core hallmark of cancer'. A balanced case.

D. C. Wallace, 'Mitochondria and cancer', *Nature Reviews Cancer* 12 (2012) 685–698. I neglected to say much about Doug Wallace's work on cancer in this chapter, in part because I discussed his ideas at some length in Chapter 6. I hardly need to say that he was one of the strongest voices calling attention to the critical role of mitochondria in cancer.

An ancient switch

P. W. Hochachka and K. B. Storey, 'Metabolic consequences of diving in animals and man', *Science* 187 (1975) 613–621. How diving animals survive long periods without oxygen – calling attention to the importance of fumarate and succinate in anaerobic metabolism. A visionary piece from the grand master of comparative biochemistry and one of his most distinguished students.

D. G. Ryan, M. P. Murphy, C. Frezza, H. A. Prag, E. T. Chouchani, L. A. O'Neill and E. L. Mills, 'Coupling Krebs cycle metabolites to signalling in immunity and cancer', *Nature Metabolism* 1 (2019) 16–33. Regrettably, I barely touched on the immune system in this book. But Krebs-cycle intermediates and their derivatives, notably itaconate, are also beginning to take centre stage in immune modulation, as outlined in this paper and several others by Evanna Mills and Luke O'Neill. (I can't resist mentioning this one for its title: E. L. Mills, B. Kelly and L. A. J. O'Neill, 'Mitochondria are the powerhouses of immunity', *Nature Immunology* 18 (2017) 488 –498.)

C. Frezza and E. Gottlieb, 'Mitochondria in cancer: not just innocent bystanders', *Seminars in Cancer Biology* 19 (2009) 4–11. A fine historically grounded overview of the Warburg effect, with a balanced appraisal of the evidence. Like Craig Thompson, Christian Frezza was here already stressing the growth requirements of cancer – Krebs-cycle intermediates are needed as precursors for biosynthesis more than they are for providing energy.

C. Frezza, 'Metabolism and cancer: the future is now', *British Journal of Cancer* 122 (2020) 133–135. A call to arms from Christian Frezza. Time the metabolic basis of cancer became mainstream.

Hypoxia in cancer

E. T. Chouchani, V. R. Pell, E. Gaude *et al.*, 'Ischaemic accumulation of succinate controls reperfusion injury through mitochondrial ROS', *Nature* 515 (2014) 431–435. A barnstorming paper that finally makes sense of reperfusion injury, from heart attacks to organ transplants. Not everyone agrees, but this is the kind of paper that reframes the problem for future generations. It certainly made sense of my own PhD, twenty years earlier.

I. H. Jain, L. Zazzeron, R. Goli *et al.*, 'Hypoxia as a therapy for mitochondrial disease', *Science* 352 (2016) 54–61. A revelatory paper from Vamsi Mootha and colleagues. Strong evidence that reactive oxygen species really do cause problems in damaged mitochondria, which can be offset by living in a 'low-oxygen tent' equivalent to Everest base camp. It works in mice but is sadly not easily achieved in humans.

Glutamine in cancer

H. Eagle, 'Nutrition needs of mammalian cells in tissue culture', *Science* 122 (1955) 501–514. One of the first papers to point to the addiction of cancer cells to glutamine, in this case in HeLa cells. On that count, I feel obliged to point you to an important book by Rebecca Skloot, *The Immortal Life of Henrietta Lacks* (Pan, 2011), after whom HeLa cells were named, which calls attention to a historical wrong.

S. L. Colombo, M. Palacios-Callender, N. Frakich, S. Carcamo, I. Kovacs, S. Tudzarova and S. Moncada, 'Molecular basis for the differential use of glucose and glutamine in cell proliferation as revealed by synchronized HeLa cells', *Proceedings of the National Academy of Sciences USA* 108 (2011) 21069–21074. Proof that cancer cells can still grow without glucose but won't if deprived of glutamine. They need mitochondria. HeLa cells again.

S. Ochoa, 'Isocitrate dehydrogenase and carbon dioxide fixation', *Journal of Biological Chemistry* 159 (1945) 243–244. A paper for the ages from the great Spanish biochemist Severo Ochoa, showing that animals can fix CO_2 by reversing bits of the Krebs cycle.

A. Mullen, W. Wheaton, E. Jin *et al.*, 'Reductive carboxylation supports growth in tumour cells with defective mitochondria', *Nature* 481 (2012) 385–388. A major paper which showed for the first time that parts of the Krebs cycle can operate in reverse in cancer, fixing CO_2 to generate citrate from α-ketoglutarate. Should have been less of a revelation, given that CO_2 fixation in animals and the reverse Krebs cycle were not exactly news. Critically, this only occurred in tumour cells with defective mitochondria, preventing normal forwards flux.

S. M. Fendt, E. L. Bell, M. A. Keibler, B. A. Olenchock, J. R. Mayers, T. M. Wasylenko, N. I. Vokes, L. Guarente, M. G. Vander Heiden and G. Stephanopoulos, 'Reductive glutamine metabolism is a function of the α-ketoglutarate to citrate ratio in cells', *Nature Communications* 4 (2013) 2236. Ratios matter in biochemistry. Which way will the Krebs cycle operate? You can predict on the basis of how much α-ketoglutarate there is relative to citrate.

A. R. Mullen, Z. Hu, X. Shi, L. Jiang, L. K. Boroughs, Z. Kovacs, R. Boriack, D. Rakheja, L. B. Sullivan, W. M. Linehan, N. S. Chandel and R. J. DeBerardinis, 'Oxidation of alpha-ketoglutarate is required for reductive carboxylation in cancer cells with mitochondrial defects', *Cell Reports* 7 (2014) 1679–1690. Complications! To convert α-ketoglutarate to citrate requires oxidising α-ketoglutarate, the exact opposite. In other words, there has to be branched flux through the Krebs cycle. Time to forget the idea of a 'normal' Krebs cycle in cancer, or life more generally.

NADPH-linked power drivers

L. A. Sazanov and J. B. Jackson, 'Proton-translocating transhydrogenase and NAD- and NADP-linked isocitrate dehydrogenases operate in a substrate cycle which contributes to fine regulation of the tricarboxylic acid cycle activity in mitochondria', *FEBS Letters* 344 (1994) 109–116. A lovely paper showing how the Krebs cycle can be turbocharged.

M. Wagner, E. Bertero, A. Nickel, M. Kohlhaas, G. E. Gibson, W. Heggermont, S. Heymans and C. Maack, 'Selective NADH communication from α-ketoglutarate dehydrogenase to mitochondrial transhydrogenase prevents reactive oxygen species formation under reducing conditions in the heart', *Basic Research in Cardiology* 115 (2020) 53. When NADH builds up it is converted into NADPH, which lowers membrane potential and regenerates the antioxidant glutathione. A good short-term use of NADH (but there can be too much of a good thing).

Sirtuins and epigenetic shifts

S.-i. Imai and L. Guarente, 'It takes two to tango: NAD$^+$ and sirtuins in aging/longevity control', *Aging and Mechanisms of Disease* 2 (2016) 16017. A nice update on how sirtuins control gene expression in relation to NAD$^+$ levels. Challenges the notion that the ratio of NAD$^+$ to NADH matters, although I'm suspicious of this conclusion in relation to the low-calorie diets that tend to activate sirtuins.

M. S. Bonkowski and D. A. Sinclair, 'Slowing ageing by design: the rise of NAD$^+$ and sirtuin-activating compounds', *Nature Reviews Molecular Cell Biology* 17 (2016) 679–690. An excellent detailed overview of the science behind the hot topic of NAD$^+$ precursors as activators of sirtuins to extend lifespan. It definitely works in flies and mice, with mild side-effects on fertility. I confess to some scepticism (I suspect a downside), but Sinclair makes a strong case for revolution in the field in his recent book *Lifespan: Why We Age – And Why We Don't Have To* (London, Thorsons, 2019).

D. V. Titov, V. Cracan, R. P. Goodman, J. Peng, Z. Grabarek and V. K. Mootha, 'Complementation of mitochondrial electron transport chain by manipulation of the NAD$^+$/NADH ratio', *Science* 352 (2016) 231–235. The activation of sirtuins depends on the NAD$^+$/NADH ratio (probably!) which in turn depends on the ability of mitochondria to oxidise NADH back to NAD$^+$. Here the redoubtable Vamsi Mootha and colleagues find clever ways of manipulating the ratio.

Respiring with glycerol phosphate

A. E. McDonald, N. Pichaud and C. A. Darveau. '"Alternative" fuels contributing to mitochondrial electron transport: Importance of non-classical pathways in the diversity of animal metabolism', *Comparative Biochemistry and Physiology B: Biochemistry & Molecular Biology* 224 (2018) 185–194. An excellent corrective to the pervasive idea that glucose is broken down to pyruvate and fed into the Krebs cycle. I've talked about a few of these alternative electron inputs to the respiratory chain in this book. If you want diagrams, check out this paper.

Erich Gnaiger, *Mitochondrial Pathways and Respiratory Control. An Introduction to OXPHOS Analysis* (Innsbruck, Bioenergetics Communications, 2020). Available here: https://doi:10.26124/bec:2020-0002. The 'bible' of fluo-respirometry, privately published by Erich Gnaiger in the tradition of Peter Mitchell's 'little grey books'; this is the little blue book. Gives practical insights into how the Krebs cycle really works. Introduces the idea of the Q junction, where electrons funnel from many substrates, including glycerol phosphate outside the mitochondria, into complex III.

Glutamine cycle in muscles

R. DeBerardinis and T. Cheng, 'Q's next: the diverse functions of glutamine in metabolism, cell biology and cancer', *Oncogene* 29 (2010) 313–324. Puts forward the disturbing notion that cancers 'deliberately' release the ammonia from glutamine, which provokes the breakdown of distant muscles, ramping up the delivery of ever more glutamine to the cancer – strip-mining the body. Like society's addiction to fossil fuels.

Ketogenic diet and cancer

D. D. Weber, S. Aminzadeh-Gohari, J. Tulipan, L. Catalano, R. G. Feichtinger and B. Kofler, 'Ketogenic diet in the treatment of cancer – where do we stand?', *Molecular Metabolism* 33 (2020) 102–121. A good balanced analysis. I've not said much about ketogenic diets in this book, but they force us to use our mitochondria. There could be many benefits, including as an adjuvant cancer treatment. I would, if I didn't enjoy my carbs so much.

Chapter 6: The flux capacitor

Evolutionary ideas of ageing

Peter Medawar, *An Unsolved Problem of Biology* (London, H. K. Lewis, 1952). The book from his inaugural lecture at UCL, which put forward a new theory of ageing that is still more or less current today.

T. Niccoli and L. Partridge, 'Ageing as a risk factor for disease', *Current Biology* 22 (2012) R741–R752. Dame Linda Partridge has been one of the most authoritative voices in the ageing field for decades, and recently stepped down from running the Institute of Healthy Ageing at UCL. Fix ageing and we can cure diseases; treat disease without considering the underlying reasons for ageing and we end up with a terrible burden of morbidity.

George C. Williams, *Adaptation and Natural Selection: A Critique of Some Current Evolutionary Thought* (Princeton, Princeton University Press, 1966). Williams independently put forward a similar view of ageing to Medawar's, known forbiddingly as antagonistic pleiotropy (meaning that genes can have positive benefits in youth that outweigh disadvantages with older age). While this book does not focus on ageing it is one of the finest and most influential pieces of evolutionary thinking of the twentieth century, and should be read by all.

André Klarsfeld and Frédéric Revah, *The Biology of Death: Origins of Mortality* (Ithaca, NY, Cornell University Press, 2000). A delightful romp through the strangeness of ageing and death in nature. Rollicking natural history with a theoretical underpinning.

Genome-wide association studies (GWAS)

Carl Zimmer, *She Has Her Mother's Laugh: The Powers, Perversions and Potential of Heredity* (London, Penguin Random House, 2018). A magisterial volume on the history and future of genetics, with excellent coverage of its murky past grounded in the warped ideas of eugenics. Brings his subject right up to date with an excellent overview of GWAS and why there might be millions of variants at play (but I still think the discipline overlooks mitochondria).

V. Tam, N. Patel, M. Turcotte, Y. Bossé, G. Paré and D. Meyre, 'Benefits and limitations of genome-wide association studies', *Nature Reviews Genetics* 20 (2019) 467–484. A recent comprehensive review of GWAS that does a good job of addressing pervasive criticisms in the literature.

Missing heritability

T. A. Manolio, F. S. Collins, N. J. Cox *et al.*, 'Finding the missing heritability of complex diseases', *Nature* 461 (2009) 747–753. One of the first papers to spell out the problem which has only partially been solved since then. For me the bottom line is most GWAS studies neglect mitochondrial DNA, no trivial omission.

Adam Rutherford, *A Brief History of Everyone Who Ever Lived: The Stories in Our Genes* (London, Weidenfeld & Nicolson, 2017). Adam Rutherford takes up the mantle of Steve Jones for the rising generation, finding his range in this sprawling, opinionated and cogent take on human genetics. Well worth reading. His more recent book *How to Argue with a Racist* (Weidenfeld & Nicolson, 2020) is even better, and should be read by everybody.

G. Pesole, J. F. Allen, N. Lane, W. Martin, D. M. Rand, G. Schatz and C. Saccone, 'The neglected genome', *EMBO Reports* 13 (2012) 473–474. A short and heartfelt plea to the GWAS community to stop discarding mitochondrial DNA.

Doug Wallace and advances in mitochondrial research

D. C. Wallace, Y. Pollack, C. L. Bunn and J. M. Eisenstadt, 'Cytoplasmic inheritance in mammalian tissue culture cells', *In Vitro* 12 (1976) 758–776. The paper from Doug Wallace's PhD thesis on cybrids treated with antibiotics.

R. E . Giles, H. Blanc, H. M. Cann and D. C. Wallace, 'Maternal inheritance of human mitochondrial DNA', *Proceedings of the National Academy of Sciences USA* 77 (1980) 6715–671. The first demonstration that mitochondrial DNA is inherited from the mother in humans, although this had already been established for other species. Maternal inheritance of mitochondria now has the status of general rule of thumb (there are no strict laws in biology).

D. A. Merriwether, A. G. Clark, S. W. Ballinger, T. G. Schurr, H. Soodyall, T. Jenkins, S. T. Sherry and D. C. Wallace, 'The structure of human mitochondrial DNA variation', *Journal of Molecular*

Evolution 33 (1991) 543–55. One of the earlier papers tracing human ancestry and migrations using mitochondrial DNA. Unlike others, Wallace never thought of mitochondrial DNA as 'neutral' (only weakly subject to selection), but perceives it as facilitating adaptation to new environments; on the flip side, mutations can also cause disease.

D. C. Wallace, M. D. Brown and M. T. Lott, 'Mitochondrial DNA variation in human evolution and disease', *Gene* 238 (1999) 211–230. A later contribution from Wallace and colleagues, taking mitochondrial diseases into more explicit consideration.

D. C. Wallace, 'Mitochondrial diseases in man and mouse', *Science* 283 (1999) 1482–1488. A seminal paper on mitochondrial diseases, comparing mouse models with human conditions. Wallace in his pomp.

W. Fan, K. G. Waymire, N. Narula, P. Li, C. Rocher, P. E. Coskun, M. A. Vannan, J. Narula, G. R. Macgregor and D. C. Wallace, 'A mouse model of mitochondrial disease reveals germline selection against severe mtDNA mutations', *Science* 319 (2008) 958–962. An important paper that reconciled a paradox in Wallace's thinking – how could a fast mitochondrial mutation rate facilitate adaptation to new environments without generating a terrible toll in mitochondrial disease? The answer was selective elimination of the worst mitochondrial mutations in the female germline. Others published similar results at the same time.

N. Lane, 'Biodiversity: on the origin of bar codes', *Nature* 462 (2009) 272–274. My own attempt to reconcile adaptation, mitochondrial mutations and the surprisingly sharp distinctions between species (so-called bar-codes in mitochondrial DNA), discussing Wallace's work.

M. Colnaghi, A. Pomiankowski and N. Lane, 'The need for high-quality oocyte mitochondria at extreme ploidy dictates mammalian germline development', *eLife* 10 (2021) e69344. A mathematical model showing that the architecture of the female germline can be understood in terms of selection against detrimental mutations in mitochondrial DNA.

D. C. Wallace, 'Mitochondria as chi', *Genetics* 179 (2008) 727–735. Pushing the boundaries of Western medicine, with its twin pillars grounded in the work of Vesalius and Mendel. An important synthesis; should be required reading for medical students.

Mitochondria in flies

D. E. Miller, K. R. Cook and R. S. Hawley, 'The joy of balancers', *PLoS Genetics* 15 (2019) e1008421. A good historical take on balancer chromosomes in *Drosophila*, along with a summary of some sticky issues.

P. Innocenti, E. H. Morrow and D. K. Dowling, 'Experimental evidence supports a sex-specific selective sieve in mitochondrial genome evolution', *Science* 332 (2011) 845–848. A seminal paper from Damian Dowling's group, showing that maternal inheritance of mitochondria really does cause problems in males.

M. F. Camus, D. J. Clancy and D. K. Dowling, 'Mitochondria, maternal inheritance, and male aging', *Current Biology* 22 (2012) 1717–1721. More from Damian Dowling, this time with my UCL colleague Flo Camus, showing that the maternal inheritance of mitochondria might even account for the shorter lifespan of men (all part of 'mother's curse').

A. L. Radzvilavicius, N. Lane and A. Pomiankowski, 'Sexual conflict explains the extraordinary diversity of mechanisms regulating mitochondrial inheritance', *BMC Biology* 15 (2017) 94. Our modelling take on why there are so many different ways of enforcing maternal inheritance of mitochondria. It boils down to an evolutionary wrestling match over who controls the destruction of male mitochondria in the sperm or the egg – the mother or the father? Or both.

Hybid breakdown does not vary with genetic distance

M. F. Camus, M. O'Leary, M. Reuter and N. Lane, 'Impact of

mitonuclear interactions on life-history responses to diet', *Philosophical Transactions of the Royal Society B* 375 (2020) 20190416C. The most striking finding here is that there is no relation between genetic distance (the number of SNPs in mitochondrial DNA) and the severity of phenotype (fertility or longevity) in crosses between different populations. 'Race' holds no meaning.

L. Carnegie, M. Reuter, K. Fowler, N. Lane and M. F. Camus, 'Mother's curse is pervasive across a large mitonuclear *Drosophila* panel', *Evolution Letters* 5 (2021) 230–239. A large panel of 81 different fly lines, in which the mitochondria are systematically mismatched against the nuclear background, demonstrates that mother's curse is real.

Redox stress caused by the antioxidant N-acetyl cysteine (NAC)

E. Rodríguez, F. Grover Thomas, M. F. Camus and N. Lane, 'Mitonuclear interactions produce diverging responses to mild stress in *Drosophila* larvae', *Frontiers in Genetics* 12 (2021) 734255. Larvae just eat and grow, placing the mitochondria under different stress compared with adult flies. Giving them the antioxidant NAC caused problems in some lines but not others, depending on the mitochondrial DNA.

M. F. Camus, W. Kotiadis, H. Carter, E. Rodriguez and N. Lane, 'Mitonuclear interactions produce extreme differences in response to redox stress', available as a preprint on *BioRxiv* (https://doi.org/10.1101/2022.02.10.479862). The paper I dwelt on at length in this chapter. A shock to me. Flies keep their ROS flux (the rate of hydrogen peroxide production) under tight homeostatic control by suppressing respiration at complex I, to the point of death. Big differences between males and females as well as different fly lines (all with equivalent genes in the nucleus) depended only on mitochondrial DNA.

Glutathione, redox stress and S-glutathionylation

P. Korge, G. Calmettes and J. N. Weiss, 'Increased reactive oxygen species production during reductive stress: the roles of mitochondrial glutathione and thioredoxin reductase', *Biochimica Biophysica Acta* 1847 (2015) 514–525. Glutathione is considered the 'master antioxidant' but there can be too much of a good thing. The problem here is called 'reductive stress', where there are (in effect) too many electrons to stay in balance. They end up dribbling onto oxygen to form reactive oxygen species.

R. J. Mailloux and W. G. Willmore, 'S-Glutathionlylation reactions in mitochondrial function and disease', *Frontiers in Cell and Developmental Biology* 2 (2014) 68. A nice review of the redox regulatory system whereby oxidised glutathione binds to proteins (*S*-glutathionylation), which can suppress respiration.

R. J. Mailloux, 'Protein *S*-glutathionylation reactions as a global inhibitor of cell metabolism for the desensitization of hydrogen peroxide signals', *Redox Biology* 32 (2020) 101472. Ryan Mailloux shows that *S*-glutathionylation suppresses hydrogen peroxide signalling – by suppressing all metabolism. Quite a mallet.

Number of heartbeats

Raymond Pearl, *The Rate of Living. Being an Account of some Experimental Studies on the Biology of Life Duration* (London, University of London Press, 1928). Raymond Pearl's rate of living theory fell out of fashion because there seemed to be many exceptions; birds, for example, live much longer than their metabolic rate would predict. But most of these exceptions can be explained simply enough. There's still truth to the idea.

Geoffrey West, *Scale: The Universal Laws of Life and Death in Organisms, Cities and Companies* (London, Weidenfeld & Nicolson, 2017). West is a physicist who has published revolutionary papers on fractal scaling, with Brian Enquist and James Brown, which I'm not citing individually here. There are plenty of objections, but the ideas are stimulating. Here, West develops them at book length, drawing parallels with companies and cities that I haven't seen anyone do since the great polymath J. B. S. Haldane.

Antioxidants don't work

J. M. Gutteridge and B. Halliwell, 'Free radicals and antioxidants in the year 2000. A historical look to the future', *Annals of the New York Academy of Sciences* 899 (2000) 136–147. The gurus of free radical biology, and authors of a famous textbook (*Free Radicals in Biology and Medicine* (Oxford University Press, 2015)). I'm citing this paper for its brevity, and the memorable line 'By the 1990s it was clear that antioxidants are not a panacea for ageing and disease, and only fringe medicine still peddles this notion.' How long does it take popular culture to catch up?

M. W. Moyer, 'The myth of antioxidants', *Scientific American* 308 (2013) 62–67. A clear presentation showing that antioxidants don't work, in either animal models or humans. Large human trials show that, if anything, antioxidant supplementation is associated with worse outcomes and a slightly higher risk of death.

M. P. Murphy, A. Holmgren, N. G. Larsson, B. Halliwell, C. J. Chang, B. Kalyanaraman, S. G. Rhee, P. J. Thornalley, L. Partridge, D. Gems, T. Nyström, V. Belousov, P. T. Schumacker and C. C. Winterbourn, 'Unraveling the biological roles of reactive oxygen species', *Cell Metabolism* 13 (2011) 361–366. A bit of a 'Hollywood' line-up, aiming for a consensus on the role of free radicals, following a meeting in Stockholm. A tall order, but worth a shot; ended up mostly with committee-speak words of caution, though nonetheless true for that.

N. Lane, 'A unifying view of ageing and disease: the double-agent theory', *Journal of Theoretical Biology* 225 (2003) 531–540. A hypothesis sketched out for the general reader in *Oxygen* (Oxford University Press, 2002) and developed more formally here. I attempt to reconcile the paradox that (were both statements true) free radicals drive ageing but antioxidants do not work. I focused on the immune system then, and would focus more on respiratory suppression now; but the bottom line is that ROS flux is controlled within tight homeostatic limits, and the body goes out of its way to prevent antioxidants interfering with those settings.

Hyperfunction

M. V. Blagosklonny, 'Aging is not programmed', *Cell Cycle* 12 (2013) 3736–3742. A clarification of his own influential thinking. Ageing is driven by 'quasi-programs' that keep on running too long; sometimes called the 'bloated soma theory'. According to this theory there is no program 'for ageing'; rather ageing results from the overrun of other developmental programs that selection is too weak, at older age, to shut down. I explain the same observations in terms of respiratory suppression producing a shift in metabolic flux and epigenetic state.

D. Gems, 'The aging–disease false dichotomy: understanding senescence as pathology', *Frontiers in Genetics* 6 (2015) 12. David Gems rails against the idea that ageing is somehow normal and 'healthy', whereas disease is pathological. The best way to treat all age-related diseases is clearly to treat the underlying problem – ageing itself. To define ageing as 'healthy' dissolves both the problem and the solution. Gems is a proponent of quasi-programs driving ageing.

Glycation damage

M. Fournet, F. Bonté and A. Desmoulière, 'Glycation damage: a possible hub for major pathophysiological disorders and aging', *Aging and Disease* 9 (2018) 880–900. A good review of the targets of glycation (the decoration of proteins, lipids and DNA with sticky, sugary tails) with age.

Variations in mitochondrial proteome between tissues

S. E. Calvo and V. K. Mootha, 'The mitochondrial proteome and human disease', *Annual Review of Genomics and Human Genetics* 11 (2010) 25–44. How much do the proteins in mitochondria differ from one tissue to another? More than you might think: nearly half of mitochondrial proteins change from tissue to tissue.

Insulin secretion depends on membrane potential

E. Heart, R. F. Corkey, J. D. Wikstrom, O. S. Shirihai and B. E. Corkey, 'Glucose-dependent increase in mitochondrial membrane potential, but not cytoplasmic calcium, correlates with insulin secretion in single islet cells', *American Journal of Physiology: Endocrinology and Metabolism* 290 (2006) E143–E148. A revelatory paper. Assuming causality … more glucose produces a higher mitochondrial membrane potential which triggers the secretion of more insulin. Is it really so simple? I love beautiful simplicity so I hope so.

A. A. Gerencser, S. A. Mookerjee, M. Jastroch and M. D. Brand, 'Positive feedback amplified the response of mitochondrial membrane potential to glucose concentration in clonal pancreatic beta cells', *Biochimica Biophysica Acta: Molecular Basis of Disease* 1863 (2017) 1054–1065. More details on how glucose modulates mitochondrial membrane potential and insulin secretion from a crack group of experts.

Links between diabetes and Alzheimer's disease

S. M. de la Monte and J. R. Wands, 'Alzheimer's disease is type 3 diabetes – evidence reviewed', *Journal of Diabetes Science and Technology* 2 (2008) 1101–1113. The evidence is mounting that Alzheimer's disease is a form of diabetes. I need to mind my diet.

P. I. Moreira, 'Sweet mitochondria: a shortcut to Alzheimer's disease', *Journal of Alzheimer's Disease* 62 (2018) 1391–1401. Never was there a more sinister use of the term 'sweet mitochondria' (the term 'diabetes mellitus' means sweet urine). Paula Moreira discusses how diabetes-linked defects in mitochondria increase the risk of Alzheimer's disease.

D. A. Butterfield and B. Halliwell, 'Oxidative stress, dysfunctional glucose metabolism and Alzheimer disease', *Nature Reviews Neuroscience* 20 (2019) 148–160. If even the redoubtable Barry Halliwell agrees, there is little room for doubt. As an aside, science is based on trust: when on unfamiliar ground, first check what somebody we trust has to say about it. I've noticed with COVID that this is a serious problem for the wider public, who can't easily turn to trusted colleagues for balanced judgement. Scientists need to do better to fix this.

Further reading

Mitochondrial-associated membranes and Alzheimer's disease

E. Area-Gomez and E. A. Schon, 'On the pathogenesis of Alzheimer's disease: the MAM hypothesis', *FASEB Journal* 31 (2017) 864–867. An extremely coherent explanation for Alzheimer's disease, linking mitochondrial damage in diabetes with calcium overload and aberrant lipid and protein processing in MAMs (mitochondria-associated membranes). Everything adds up.

E. Area-Gomez, C. Guardia-Laguarta, E. A. Schon and S. Przedborski, 'Mitochondria, OxPhos, and neurodegeneration: cells are not just running out of gas', *Journal of Clinical Investigation* 129 (2019) 34–45. My favourite paper on the MAM hypothesis. Not just a coherent explanation for Alzheimer's disease, but also why more conventional mitochondria-centric explanations fall short. I don't understand why the idea seems to have lukewarm support in the mainstream field. Too many competing hypotheses? This is one of the best.

Powering up mitochondria by calcium activation of pyruvate dehydrogenase

A. P. Wescott, J. P. Y. Kao, W. J. Lederer and L. Boyman, 'Voltage-energized calcium-sensitive ATP production by mitochondria', *Nature Metabolism* 1 (2019) 975–984. Just a wonderful paper. Calcium floods into the mitochondria (from the MAMs, though the authors don't explicitly make the link) and ramps up the activity of pyruvate dehydrogenase, which cranks up the Krebs cycle, mitochondrial membrane potential and ATP synthesis. The membrane potential changes quickly and can achieve the highest levels only in this way.

Biology by the numbers

Ron Milo and Rob Phillips, *Cell Biology by the Numbers* (New York, Garland Science, 2016). Engaging, quantitative answers to all the questions you never knew you had about cells. I'm citing this book in relation to Joana Xavier's calculations on the number of metabolic reactions per second in cells. They don't calculate this particular parameter, but they calculate many other amazing things that you really should know about.

Epilogue: Self

Anaesthetics and mitochondria

L. Turin, E. M. C. Skoulakis and A. P. Horsfield, 'Electron spin changes during general anesthesia in *Drosophila*', *Proceedings of the National Academy of Sciences USA* 111 (2014) E3524–E3533. Luca Turin and colleagues on how general anaesthetics influence oxygen and respiration in mitochondria. An excellent paper in a prestigious journal, but they don't quite crack the problem here ...

L. Turin and E. M. C. Skoulakis 'Electron spin resonance (EPR) in *Drosophila* and general anesthesia', *Methods in Enzymology* 603 (2018) 115–128 ... But they do in this paper, albeit in a less visible place. A couple of clever tricks allow them to show that general anaesthetics short-circuit the flow of electrons to oxygen in respiration. They've not nailed causality, but for the first time (to my mind) they have opened a path to a clear mechanistic explanation of consciousness.

A. Gaitanidis, A. Sotgui and L. Turin, 'Spontaneous radiofrequency emission from electron spins within *Drosophila*: a novel biological signal', *arXiv*:1907.04764 (2019). A tough paper for anyone who is not a biophysicist, but this is radical material that is opening the doors to real twenty-first-century science.

EEG and mitochondria

M. X. Cohen, 'Where does EEG come from and what does it mean?', *Trends in Neurosciences* 40 (2017) 208–218. Well worth a read. The fact that we've known about the electroencephalogram for a century, and can measure changes with great precision in sleep and many disease states, should not blind you to the embarrassing fact that we have no clear idea of exactly what generates it. Actually, it shouldn't be embarrassing. Sing it from the rooftops! Science is about what we don't know; the fact we don't understand the EEG means there's exciting work to be done. Never be fooled into thinking we know most of the answers.

Further reading

T. Yardeni, A. G. Cristancho, A. J. McCoy, P. M. Schaefer, M. J. McManus, E. D. Marsh and D. C. Wallace, 'An mtDNA mutant mouse demonstrates that mitochondrial deficiency can result in autism endophenotypes', *Proceedings of the National Academy of Sciences USA* 118 (2021) e2021429118. A specific mitochondrial mutation in mice leads to behavioural changes that are reminiscent of autism in humans; and strikingly, the EEG also shows patterns that are linked with autism in humans. At least consistent with the much bolder hypothesis that mitochondrial membranes *generate* the EEG.

Electrical fields in development

M. Levin and C. J. Martyniuk, 'The bioelectric code: an ancient computational medium for dynamic control of growth and form', *Biosystems* 164 (2018) 76–93. Michael Levin is currently doing some of the most exciting work in biology. Electrical fields determine development; genes are secondary. So far that mostly goes for flatworms, but could it be more general? I think so.

M. Levin and D. C. Dennett, 'Cognition all the way down', *Aeon*, 13 October 2020. Here Levin teams up with Daniel Dennett, who has a very acute nose for what is important in biology, especially in relation to consciousness. We are about to make serious progress on the most exciting and important open question in biology.

D. Ren, Z. Nemati, C. H. Lee, J. Li, K. Haddadi, D. C. Wallace and P. J. Burke, 'An ultra-high bandwidth nano-electronic interface to the interior of living cells with integrated fluorescence readout of metabolic activity', *Scientific Reports* 10 (2020) 10756. Doug Wallace is still at it, teaming up with a group of nanotechnologists on the vibrant campus in Philadelphia to measure electrical fields from mitochondria in single cells. The first steps towards addressing a tough question.

Consciousness and self

Derek Denton, *The Primordial Emotions. The Dawning of*

Consciousness (Oxford, Oxford University Press, 2006). Denton worked for many years on salt balance and thirst, which led him to explore the 'imperious states of arousal and compelling intentions to act' found across the animal kingdom, focusing his attention on the ancient parts of the brainstem that we share with most animals. Now in his nineties, Denton is still active; and others, including Mark Solms and Peter Godfrey-Smith (who has written two fine books on consciousness, *Other Minds* and *Metazoa*) are beginning to argue along similar lines.

Mark Solms, *The Hidden Spring: A Journey to the Source of Consciousness* (London, Profile Books, and New York, W. W. Norton, 2021). I was going to discuss Solms's recent ideas in my Epilogue but they didn't quite fit. Where he writes about the central nervous system, I think about cells. Solms develops the idea that organisms strive to lower their free energy, which is roughly their level of physiological discomfort in relation to the environment; an attempt to restore homeostasis. Behaviour aims to make biochemistry more comfortable. The membrane around cells transforms signals from the outside world into the language of biochemistry, while the language of biochemistry is transformed into the electrical fields on the membrane, uniting the molecules within as a 'self'. The language of consciousness is the language of electrical fields on membranes.

M. Solms and K. Friston, 'How and why consciousness arises: some considerations from physics and physiology', *Journal of Consciousness Studies* 25 (2018) 202–238. Solms's ideas on consciousness were developed with Karl Friston, whose concept of free energy in this context is sometimes called 'Friston free energy' to distinguish it from more conventional usage. This formal paper summarises their ideas.

J. McFadden, 'Integrating information in the brain's EM field: the cemi field theory of consciousness', *Neuroscience of Consciousness* 2020 (2020) niaao16. Johnjoe McFadden has been thinking about electrical fields in relation to consciousness for several decades, and this is a recent update on his cemi field theory. I need to talk to him about mitochondria ...

Further reading

Membrane potential in bacteria and cell death

E. S. Lander, 'The heroes of CRISPR', *Cell* 164 (2016) 18–28. I mentioned CRISPR (now widely used for clever gene editing) in relation to the bacterial immune system. This paper covers some of the backstory, tracing several decades of curiosity-driven research on bacterial ecosystems. Many transformative breakthroughs in science come from unexpected quarters, and research done with no utility in mind. We should keep funding it.

D. Refardt, T. Bergmiller and R. Kümmerli, 'Altruism can evolve when relatedness is low: evidence from bacteria committing suicide upon phage infection', *Proceedings of the Royal Society B* 280 (2013) 20123035. This paper is really about kin selection in bacteria, proving it is worth their while to kill themselves to save barely related bacteria, if they are infected by a virus that will kill them anyway. I was struck by how bacteria kill themselves – they collapse their membrane potential, dying in seconds, and preventing the virus taking over. Death is the permanent loss of membrane potential; plainly that kills the conscious mind too.

H. Strahl and L. W. Hamoen, 'Membrane potential is important for bacterial cell division', *Proceedings of the National Academy of Sciences USA* 107 (2010) 12281–12286. How do bacteria divide down the middle to form two daughter cells? It turns out that oscillator proteins move rapidly from one end of the cell to the other and eventually locate the midpoint, where they serve as a scaffold for the constricting apparatus of cell division (the Z ring). What came as a revelation is that they can only achieve this when there is an electrical membrane potential – collapse the potential and bacteria can't figure out how to divide in two. I suspect that this loss of orientation is just one manifestation of how electrical fields give cells their unity.

ACKNOWLEDGEMENTS

This book opens with a description of an empty city, lacking the hurly burly of people and energy flow. Little did I realise when writing that passage that it would require such a trivial leap of the imagination – I had written it before Covid-19 sucked the life out of cities the world over. Rather than rewriting my opening in light of Covid, I decided to leave it unchanged, as my point had nothing to do with Covid, and I hope this book will outlast the pandemic. Nonetheless, much of the book was written during the Covid years, and I was deeply influenced by many long walks through the empty streets of London with my wife, Dr Ana Hidalgo-Simón, and our two boys Eneko and Hugo, often lost in discussion of the ideas presented here. I habitually thank Ana last in my acknowledgements, but so much of this book found its space in those empty streets that I must now thank her first. Ana had encouraged me to write this specific book, to try to bring the enigmatic Krebs cycle to life, and it would never have happened without her. Nor would it have been half as good without those long walks and talks, and Ana's ever-vital ideas. Books evolve as they are written. This was conceived as short and light-footed. It ended up longer and weightier than intended, but I hope it has gained in coherence and has something important to say. The Krebs cycle is more central to life than even I had imagined when I started writing. That changing perspective owes much to Ana's sharp thinking and clarity. As usual, she has read every word more than once, and never hesitated to tell me the truth, however painful it was to bear. Words cannot begin to convey my debt to her.

I must thank, too, my fantastic lab group at UCL. They have

had to fight their way through Covid, with the lab closed for several months and disrupted for much longer, leaving them worried about how to finish their PhDs, or compete for post-doc positions. Thankfully, none faced personal tragedy. Somehow, they made progress through those difficult times. Their work, too, has resolved much in my mind. I have two labs, one working on the origin of life and the other on mitochondrial function in fruit flies. Before reading this book, you might have thought those subjects had little in common beyond the centrality of energy flow. Yet both groups, in their different ways, have been working on Krebs-cycle intermediates, either as a product of CO_2 fixation at the origin of life, or as the severely disturbed downstream consequences of mitonuclear incompatibilities in ageing flies. Grappling with new results while writing this book, sometimes watching cherished ideas implode, is one of the more chastening and valuable experiences of research. We can't afford to take our own ideas with more than a grain of salt. But any ideas that survive this crucible are all the better for it. The great physicist Ernest Rutherford once ruminated 'We haven't got the money, so we'll have to think.' Covid has likewise forced us to think. This book has been forged in the crucible of thinking hard about our experiments. So I thank all my close colleagues at UCL, not just those in my group, but all those who have kept me on my toes: Prof. Andrew Pomiankowski, Prof. Finn Werner, Prof. Graham Shields, Prof. John Allen, Dr Amandine Maréchal, Dr Flo Camus, Dr Seán Jordan, Dr Rafaela Vasiliadou, Dr Will Kotiadis, Dr Enrique Rodriguez, Dr Gla Inwongwan, Dr Stefano Bettinazzi, Dr Joana Xavier, Dr Feixue Liu, Stuart Harrison, Silvana Pinna, Marco Colnaghi, Raquel Nunes-Palmeria, Hanadi Rammu, Aaron Halpern, Ion Ioannou, Finley Grover Thomas, Caecilia Kunz, Toby Harries and Kaan Suman. Some of them have read chapters too and commented, for which I am very grateful. Thanks too to former students, now thriving in the outer scientific world, Dr Eloi Camprubi and Dr Victor Sojo. There are more, but I have to stop somewhere.

Many friends and colleagues have also read either chapters or the whole book. While on the theme of Covid, I want to thank in particular Diego Maria Bertini. Diego emailed me from his hospital bed in Italy, while recovering from severe Covid during that traumatic Italian spring of 2020. Lacking much to alleviate his long days of recovery, he had been reading material from my web page, and leapt at the chance to read my chapters as I wrote them. I could not have imagined how valuable his comments would be. I thank him for his unfaltering enthusiasm, his love of words, of literature and science, for his poetry, and more than once for helping me to find the right words. All this despite a painfully slow recovery from Covid (his expression 'my brain feels as if it is wrapped in a plastic bag' has haunted me). I hope we meet in person some day.

A few words of thanks to other friends, who have read chapters or the whole book and given me feedback on pitch, content and style. Mike Carter has read all my books, chapter by chapter, and never fails to keep me going through darker days with unreasonably enthusiastic comments, to say nothing of the music, conversation, sea walks and good cheer. We used to climb together, but now we reminisce. Allyson Jones has also read the whole book, pointing out more stylistic *faux pas* than I would care to admit, while celebrating the bigger picture and occasionally dredging up some details from her deeply buried degree in biochemistry. She racked her mind over possible titles. Titles are never easy, and Allyson has one of the most inventive minds I know. The fact that she didn't better the current title put my mind at rest. I am indebted to the legendary film editor Walter Murch, who is the only person I know who uses scientific metaphors to illuminate points in the arts – who knew that the gene-editing spliceosome in eukaryotes could throw such light on the esoteric practices of movie editing? Walter reads widely across the sciences and is a fine writer with much to say about pace and tone. I shall never enter the sea again without thinking of his admonishments whenever the upwelling depths of biochemistry suddenly chilled my text. I can only hope that I have warmed the offending waters appropriately. Thanks as well to Emily MacKay, who fought her way through several chapters laced with chilly depths, having returned to university to take a degree in nutrition and medical sciences. She had to capitulate to the demands of serious study before finding her way back to warmer waters. And thanks to Wai Mun

Acknowledgements

Yoon, who bubbles over with his own profligate ideas, always seeing the logical implications and the way ahead.

I need to thank a few specialist colleagues and friends for reading over specific chapters. I treasure their comments, in part for confirming that I had not made any egregious errors (though they all found a few trivial mistakes), in part for pointing me in new directions, and best of all for being genuinely enthusiastic about this project. Admittedly, if I failed to enthuse those who have devoted their lives to science then what hope do I have? Yet hardened professionals also become jaded to old news, and I'd like to think my chapters were fresh enough to rekindle some youthful excitement. First, I must thank Prof. Don Braben, who has read every chapter, despite his own writing projects. Don never ceases to confront the limits of science, and the possibilities of finding new paths into the unknown. His enthusiasm is uplifting, and more infectious than Covid. I thank Prof. Mike Russell, who has had the great generosity of spirit to read and comment in detail on my chapters even when he might not agree with everything I have written. He has been inspiring to me not only in his visionary science, but also in his energy and warmth of friendship. I am indebted to Prof. Lee Sweetlove, who has read the whole book and made many penetrating comments, while celebrating the bigger picture. No one in the world today knows more about the Krebs cycle than Lee, yet he can still barely trammel his excitement about intermediary metabolism. I'll never forget our lunch together, where we lost ourselves in a six-hour discussion on why the cycle is a cycle and other abstruse questions. Rarely have science and pure pleasure aligned so closely. And I thank Prof. Christian Frezza, who some years ago made me understand viscerally how central mitochondria are to cancer, not for ATP synthesis, but for the way they shape Krebs-cycle flux and epigenetic signalling. Christian is utterly meticulous, balanced and enthusiastic. It's a sorry outcome for the UK that he has left Cambridge and moved to Cologne. I'm conscious that I'm overusing the word enthusiasm, but I find it is a common denominator among those who love the Krebs cycle.

While on the subject of cancer, I must thank Prof. Frank Sullivan, formerly of the US National Cancer Institute, and now returned to clinical practice in Galway. Frank brings a tough hands-on perspective to the difficulty of treating patients with prostate cancer, and has long seen metabolism and energy flow as critical to effective treatment. He reads widely around science and medicine, and I'm honoured that he has found my writing helpful. I thank the distinguished bioenergeticist Prof. Franklin Harold for detailed comments on much of my book. Frank is in his nineties, yet is still writing books himself, books that continue to brim with insight and poetry. We have been discussing bioenergetics and evolution by email and occasionally in person for two decades. I will never forget our pilgrimage together to Darwin's home at Down House. Frank taught me the value of open-minded scepticism in science, and he is unyielding in applying this philosophy to my writings too. But he has also allowed himself to be somewhat caught up in the ideas presented here. And I thank Prof. Mårten Wikström, veteran of the later 'Ox Phos wars', who after eight years eventually persuaded Peter Mitchell that cytochrome oxidase did indeed pump protons. Mitchell admitted having 'seen that artefact before,' once again highlighting the problems of interpretation. I thank Mårten for his stimulating discussion and valuable corrections on these questions. Thanks too to Prof. Alistair Nunn, also deeply engaged in mitochondrial research, who has read much of the book and offered stimulating comments from the viewpoint of quantum biology.

A short note on consciousness. I wrote a chapter on consciousness in *Life Ascending* in 2009, which failed to answer the question (I'm hardly alone in that). That led me into a long and enjoyable correspondence with Prof. Derek Denton, now in his nineties, yet still overflowing with energy and plans. He had been drawn into the subject through his work on salt balance and thirst in animals. I confess that I had put the question aside until Derek invited me to give the Derek Denton Oration in the Arts and Sciences in Melbourne in 2019 – on condition that I addressed consciousness, if only in brief. While wrestling with the question, I immersed myself in the revelatory work of Prof. Luca Turin (linking it in my own mind with Prof. Doug Wallace's ideas and my thoughts on the electrical unity of cells). In Australia, I was rigorously grilled over dinner for two hours on the subject by the assembled dignitaries, including Derek and the philosopher and writer Prof. Peter Godfrey Smith. I have tried to capture these thoughts in the epilogue to this book. Happily, that has renewed correspondence with Prof. Dan Dennett too, as well as Doug Wallace, Luca Turin and Michael Levin. I have the feeling we are about to make real progress. As Dan wrote, 'What interesting times!'

I would like to thank Bill Gates and his team, especially Dr Trevor Mundel and Dr Niranjan Bose, for their sustained interest in my work and their support. Rarely have I met such a well-informed and intellectually engaged group of people, who have a genuinely burning mission to improve the world. I'm proud to have contributed to their thinking about the evolutionary basis of health and medicine, if only in a small way. I hope this book gives more food for thought.

It remains for me to thank my family, especially my father, the historian Anthony Lane, who as always has read every chapter and commented on infelicities of style, signposting and historical context, despite harbouring little love of molecules. Love is the watchword here. Some find science too cold and rational for their taste, and seek meaning in religion or spirituality. I can find plenty of meaning in science, but it would be little solace without the love of my family: my parents, brother and sister-in-law in Yorkshire, and my family in Spain and Italy. I have already mentioned Ana, but here I need to thank her and our two sons for the love and meaning they bring to my life. I really could not be luckier. Despite our best efforts to inculcate a love of science in Eneko and Hugo, it's already clear that they will follow different paths through life. I look forward to seeing where they lead.

On the subject of family and meaning, I want to thank Mary Jane Ackland-Snow for allowing me to share a few lines about Ian, and to dedicate this book to his memory. Perhaps our best and most lasting legacy is to influence the lives of others for the better. Ian had an immense impact on my life. He had buoyant energy, warmth of spirit and a zestful sense that anything is possible, which would make those he touched strive to be the best we can be. I hadn't appreciated how many other people felt the same until his funeral. We will all remember him with love and admiration. Heartfelt condolences to Mary Jane.

Last but not least, I thank Caroline Dawnay of United Agents, and my publishers in London and New York, Profile Books and W. W. Norton respectively. I thank Caroline for her unwavering belief and encouragement, and for reading and commenting on the whole book as she always does. Thanks to Ed Lake at Profile Books and Brendan Curry at W. W. Norton for balancing encouragement with reality. Ed's comment that my first two chapters presented 'a pretty forbidding hurdle' to the general reader was not exactly welcome, but it forced me to rewrite those chapters extensively. When he later wrote that my revised text now had a 'a barrelling energy ... I'm thrilled with what you've done', I was not only relieved but believed him for his earlier honesty. So I thank Ed and Brendan for their love of good writing and their honesty in fostering it. Thanks too to Nick Allen, who copy-edited the book with great sensitivity and skill; to Paul Forty, for editing and producing the book with fine-grained attention to detail; to Bill Johncocks, who showed me that a good index is a work of art; to Valentina Zanca for her unbounded enthusiasm in presenting the book to the outside world; to Andrew Franklin, for creating an intellectual temple in the air at Profile; and to Andrey Kurochkin for turning my amateurish hand-drawn illustrations into molecular portraits of spellbinding beauty (and for responding patiently to my obsessively detailed corrections). Finally, I thank Cliff Hanks for bringing the poem 'Like Most Revelations' to my attention, and Richard Howard for his gracious permission to reprint the poem. It somehow captures the essence of what I have tried to say in this book, yet in a few lines from another dimension.

INDEX

Note: The index covers the main text and appendices. The suffix 'n' plus a number indicates a relevant footnote on that page. Page numbers in *italics* indicate illustrations.

Index